T0189230

Advances in Information Security

Volume 69

Series editor
Sushil Jajodia, George Mason University, Fairfax, VA, USA

More information about this series at http://www.springer.com/series/5576

Tianqing Zhu • Gang Li • Wanlei Zhou
Philip S. Yu

Differential Privacy and Applications

 Springer

Tianqing Zhu
Deakin University
Melbourne, Australia

Wanlei Zhou
Deakin University
Melbourne, Australia

Gang Li 🆔
Deakin University
Melbourne, Australia

Philip S. Yu
University of Illinois at Chicago
Chicago, IL, USA

ISSN 1568-2633
Advances in Information Security
ISBN 978-3-319-87211-7   ISBN 978-3-319-62004-6   (eBook)
DOI 10.1007/978-3-319-62004-6

Printed on acid-free paper

This Springer imprint is published by Springer Nature
The registered company is Springer International Publishing AG
The registered company address is: Gewerbestrasse 11, 6330 Cham, Switzerland

Preface

Corporations, organizations, and governments have collected, digitized, and stored information in digital forms since the invention of computers, and the speed of such data collection and volumes of stored data have increased dramatically over the past few years, thanks to the pervasiveness of computing devices and the associated applications that are closely linked to our daily lives. For example, hospitals collect records of patients, search engines collect online behaviors of users, social network sites collect connected friends of people, e-commerce sites collect shopping habits of customers, and toll road authorities collect travel details of vehicles. Such huge amounts of datasets provide excellent opportunities for businesses and governments to improve their services and to bring economic and social benefits, especially through the use of technologies dealing with big data, including data mining, machine learning, artificial intelligence, data visualization, and data analytics. For example, by releasing some statistics of hospital records may help medical research to combat diseases. However, as most of the collected datasets are personally related and contain private or sensitive information, data releasing may also provide a fertile ground for adversaries to obtain certain private and sensitive information, even though simple anonymization techniques are used to hide such information. Privacy preserving has, therefore, become an urgent issue that needs to be addressed in the digital age.

Differential privacy is a new and promising privacy framework and has become a popular research topic in both academia and industry. It is one of the most prevalent privacy models as it provides a rigorous and provable privacy notion that can be applied in various research areas, and can be potentially implemented in various application scenarios. The goal of this book is to summarize and analyze the state-of-the-art research and investigations in the field of differential privacy and its applications in privacy-preserving data publishing and releasing, so as to provide an approachable strategy for researchers and engineers to implement this new framework in real world applications.

This is the first book with a balanced view on differential privacy theory and its applications, as most existing books related to privacy preserving either do not touch the topic of differential privacy or only focus on the theoretical analysis of differential privacy. Instead of using abstract and complex notions to describe the concepts, methods, algorithms, and analysis on differential privacy, this book presents these difficult topics in a combination of applications, in order to help students, researchers, and engineers with less mathematical background understand the new concepts and framework, enabling a wider adoption and implementation of differential privacy in the real world. The striking features of the book, differs from others, can be illustrated from three basic aspects:

- A detailed coverage on differential privacy in the perspective of engineering, rather than computing theory. The most difficult part in comprehending differential privacy is the complexity and the level of abstract of the theory. This book presents the theory of differential privacy in a more natural and easy to understand way.
- A rich set of state-of-the-art examples on various application areas helps readers to understand how to implement differential privacy in real world scenarios. Each application example includes a brief introduction to the problem and its challenges, a detailed implementation of differential privacy to solve the problem, and an analysis on the privacy and utility.
- A comprehensive collection of contemporary research results and issues in differential privacy. Apart from the basic theory, most of the contents of the book are from the recent publications in the last 5 years, reflecting the state-of-the-art of research and development in the area of differential privacy.

This book intends to enable readers, especially postgraduate and senior undergraduate students, to study up-to-date concepts, methods, algorithms, and analytic skills for building modern privacy-preserving applications through differential privacy. It enables students not only to master the concepts and theories in relation to differential privacy but also to readily use the material introduced into implementation practices. Therefore, the book is divided into two parts: theory and applications. In the theory part, after an introduction of the differential privacy preliminaries, the book presents detailed descriptions from an engineering viewpoint on areas of differentially private data publishing and differentially private data analysis where research on differential privacy has been conducted. In the applications part, after a summary on the steps to follow when solving the privacy-preserving problem in a particular application, the book then presents a number of state-of-the-art application areas of differential privacy, including differentially private social network data publishing, differentially private recommender system, differential location privacy, spatial crowdsourcing with differential privacy preservation, and correlated differential privacy for non-IID datasets. The book also includes a final chapter on the future direction of differential privacy and its applications.

Acknowledgments

We are grateful to many research students and colleagues at Deakin University in Melbourne and University of Illinois at Chicago, who have made a lot of comments to our presentations as their comments inspire us to write this book. We would like to acknowledge some support from research grants we have received, in particular, the Australian Research Council Grant no. DP1095498, LP120200266, and DP140103649, NSF through grants IIS-1526499, and CNS-1626432, and NSFC (Nos. 61672313, 61502362). Some interesting research results presented in the book are taken from our research papers that indeed (partially) supported through these grants. We also would like to express our appreciations to the editors in Springer, especially Susan Lagerstrom-Fife, for the excellent professional support. Finally we are grateful to the family of each of us for their consistent and persistent supports. Without their support, the book may just become some unpublished discussions.

Melbourne, Australia	Tianqing Zhu
Melbourne, Australia	Wanlei Zhou
Melbourne, Australia	Gang Li
Chicago, IL, USA	Philip S. Yu
May 2017	

Contents

Chapter 1
Introduction

1.1 Privacy Preserving Data Publishing and Analysis

Over the past two decades, digital information collected by corporations, organizations and governments has resulted in huge number of datasets, and the speed of such data collection has increased dramatically over the last a few years. A data collector, also known as a *curator*, is in charge of publishing data for further analysis [8].

Figure 1.1 illustrates the data publishing and analysis process, in which there are roughly two stages. In the first stage, data collection stage, individuals submit their personal information to a dataset. Amount of data are collected in their relative area, such as medical data, data from banks, social network data, etc. The curator has the full management of this dataset. The second stage is the data publishing or analysis stage. Data publishing aims to share datasets or some query results to public users. In some literature, this scenario is known as data sharing or data release. Data analysis provides data models to the public users. It may be associated with some particular machine learning or data mining algorithms. Both data publishing and analysis bring social benefits, such as providing better services, publishing official statistics, providing data mining or machine learning tasks, etc.

As shown in Fig. 1.1, most of the collected datasets are personally related and contain private or sensitive information. The privacy violation may occur in both stages. If curators are not trustworthy, personal information will be disclosed directly in data collection stage. Even though curators can be trusted and apply several simple anonymization techniques, when he/she publishes aggregate information to the public, personal information may be disclosed as the public is normally not trustworthy [50, 109, 236]. Privacy-preserving has therefore become an urgent issue that needs to be addressed [84].

© Springer International Publishing AG 2017

T. Zhu et al., *Differential Privacy and Applications*,
Advances in Information Security 69, DOI 10.1007/978-3-319-62004-6_1

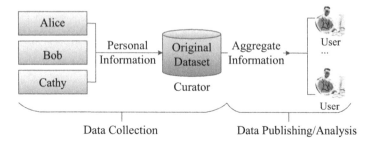

Fig. 1.1 Privacy preserving data publishing and analysis

1.2 Privacy Violations

In 1998, Latanya Sweeney provided a famous example about de anonymization on a published medical dataset [197, 211]. She shows that even with removing all explicit identifiers, individuals in the medical dataset still can be re-identified by linking with a public voter list through the combination of zip code, date of birth and gender. The information that the adversary can link with is defined as background information [211]. According to Sweeny, 87% of the U.S. population in the censorship dataset are likely to be figured out by the combination of some attributes, which is defined as Quasi-Identifiers [211]. Narayanan [167] provides another example on partly de-anoymization of Netfix dataset.

In 2007, Netflix offered a 1,000,000 prize for a 10% improvement in its recommendation system and released a training dataset with 500,000 anonymous movie ratings by subscribers. To protect the subscribers, all personal information has been removed and all subscribers' IDs have been replaced by randomly-assigned numbers. However, by linking it with the International Movie DataBase (IMDB) dataset, another online movie rating portal with users information, Narayanan et al. [166] partly de-anonymized Netflix anonymized training dataset.

With the rising of new technologies, there are more privacy violations on data publishing or analysis. For example, Mudhakar et al. [205] deanonymized mobility traces by using social networks as a side-channel. Stevens et al. [131] exploited skype, a famous P2P communications software to invade users' location and sharing information. De Montjoye [54] showed that in a dataset where the location of an individual is specified hourly, four location points are sufficient to uniquely identify 95% of the individuals.

In addition, the curators may not trustworthy. They may sell individual's personal information in purpose, or be hacked by adversaries. For example, Bose, a manufacturer of audio equipment, spied on its wireless headphone customers by using an app that tracks the music, podcasts and other audio they listen to, and violates their privacy rights by selling the information without permission [2]. Information associated with at least 500 million user accounts was stolen from Yahoo's network in 2014 by what it believed was a "state-sponsored actor" [4]. And Telstra, an

Australia telecom company, breached the privacy of 15,775 customers when their information was made publicly available on the internet between February 2012 and May 2013 [3].

All of these example's shows that simple anonymization is insufficient for privacy preserving. The adversary with background information on an individual can still has chance to identify the individual's records. People need to seek more rigorous way to guarantee personal sensitive information.

1.3 Privacy Models

Research communities have proposed various methods to preserve privacy [8, 148], and have designed a number of metrics to evaluate the privacy level of these methods. The methods and their privacy criteria are defined as the *privacy model*. To preserve privacy in the data collection stage, the privacy model is inserted between curators and individuals who submitted their personal information. The privacy model processes the personal information before submitting to the untrusted curator. To preserve privacy in the data publishing and analysis stage, the privacy model is placed between curator and public users. All data go through the privacy model should satisfy the requirements of the model. Figure 1.2 shows different placement of privacy models.

The most popular privacy model is the *k-anonymity* [197, 211], which requires that an individual should not be identifiable from a group of size smaller than *k*. Also, there are a number of other privacy models, such as *l-diversity* [151], *t-closeness* [142, 143], and *δ-presence* [169].

However, the vulnerability of these models lies in the fact that they can be easily compromised by uncontrolled background information [231]. From the published dataset, users, including the adversary, can possibly figure out the privacy requirement and anonymization operations the curator used. Wong et al. [231] show that such additional background information can lead to extra information that facilitates the privacy compromising. For example, when a dataset is generalized

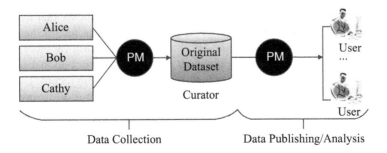

Fig. 1.2 Privacy model

until it minimally meets the *k-anonymity* requirement, the attacker would exploit this minimality *equivalence groups* to reverse the anonymization and explore the possible version of the original table. This process is called the *minimality attack* [50, 109, 236], a special method in the *record linkage* attack.

Furthermore, several other new attacks are emerging to against traditional privacy models. Composition attack [86] refers to the scenario that the adversary gleans from other channels such as the Web, public records or domain knowledge to obtain the background information. They explore how published dataset is at risk in the face of rich, realistic sources of background information. Kifer presents the deFinetti attack [125] by attacking a popular data anonymous schemes. The attacker only needs to know the insensitive attributes of an individual in the dataset, and can then carry out this attack by building a machine learning model over the sanitized dataset. The attack exploits a subtle flaw in the way that prior work computes the probability of disclosure of a sensitive attribute [125]. Moreover, Wong et al. [232] propose a new attack named the foreground knowledge attack. If dataset is not properly anonymized, patterns can be generated from the published dataset and be utilized by the adversary to breach individual privacy. This kind of attack is referred as foreground knowledge attack, on a contrary to the background information.

In general, researchers assume that the attacker has limited background information and tries to profile this information when designing privacy models, which makes these models tightly coupling with background information. However, background information is hardly to predict or model. it is impractical to know what kind of background information exactly the adversary might has. So people have few opportunity to depict or control the information that adversary has. Difficulty in modeling the background information is the inherent weakness in those traditional privacy models. Consequently, Researchers have considered a solid privacy model that is robust enough to provide a provable privacy guarantee against the background information.

1.4 Differential Privacy

There comes the *differential privacy* model, which is a solid privacy model that provides a provable privacy guarantee for individuals. It assumes that even if an adversary knows all the other records in a dataset except one record, he/she still cannot infer the information contained in that unknown record. In another word, even the adversary get to know maximum background information except the record he/she wants to know, he/she cannot identify the specific record. Under this assumption, differential privacy theoretically proves that there is a low probability of the adversary figuring out the unknown record. Compared to the previous privacy models, differential privacy can successfully resist background attack and provide a provable privacy guarantee.

The first steps toward differential privacy were taken in 2003. But it was not until 2006 that Dwork et al. discussed the technology in its present form.

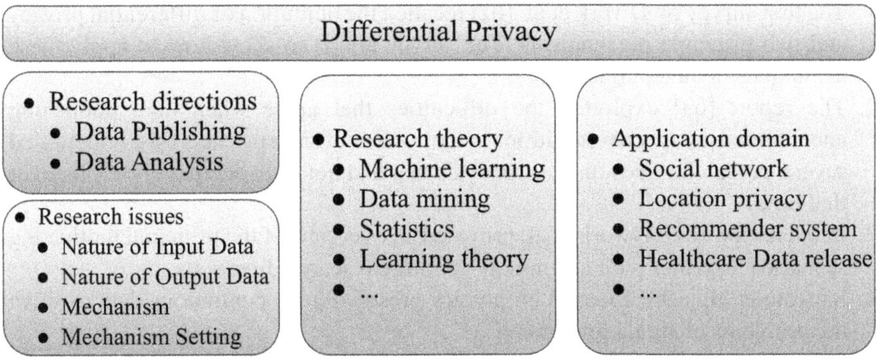

Fig. 1.3   Key components associated with differential privacy

Since then, a significant body of work has been developed by scientists around the world. Differential privacy is catching the attention of academics. In 2012, Microsoft releases a whitepaper titled *Differential Privacy for Everyone* [1], striving to translate this research into new privacy-enhancing technologies.

Differential privacy has recently been considered a promising privacy-preserving model. The interest in this area is very high and the notion is spanning in a range of research areas, ranging from the privacy community, to the data science communities including machine learning, data mining, statistics and learning theory. Figure 1.3 shows the key components associated with differential privacy research. There are roughly two major directions on differential privacy research, data publishing and data analysis. In both direction, the main research issues include the nature of input and output data, mechanism design and settings. The research methods and theories involve machine learning, data mining, statistics, and cryptograph. Much work has been conducted in a number of application domains, including social network, location privacy, recommender systems and other applications.

This book provides a structured and comprehensive overview of the extensive research on differential privacy, spanning multiple research areas and application domains. It will facilitate better understanding on areas of data publishing and data analysis in which research on differential privacy has been conducted, and how techniques developed in one area can be applied in other domains.

1.5   Outline and Book Overview

The initial work on differential privacy was pioneered by Dwork et al. [61] in 2006. The basic idea can be found in a series of works [62–64, 72, 75, 198].

1. The first survey by Dwork et al. [62] recalled the definition of differential privacy and two principle mechanisms. The survey aimed to show how to apply these techniques in data publishing.
2. The report [63] exploited the difficulties that arose when data publishing encountered prospective solutions in the context of statistics analysis. It identified several research issues on data analysis that had not been adequately explored at that time.
3. In the review [64], Dwork et al. provided an overview of the principal motivating scenarios, together with a summary of future research directions.
4. Sarwate et al. [198] focused on privacy preserving for continuous data to solve the problems of signal processing.
5. Recently, a survey of attacks on private data has been proposed Dwork et al. [75], who summarize possible attacks that compromise personal privacy.
6. A book by Dwork et al. [72] presented an accessible starting place for anyone looking to learn about the theory of differential privacy.

Those surveys and the book focus on the concepts and theories of differential privacy; however, the mathematical theories are not easily implemented into applications directly. Yet, after more than 10 years of theoretical development, a significant number of new technologies and applications have appeared in this area. We believe that now is a good time to summarize the new technologies and address the gap between theory and application.

Here we attempt to find a clearer way to present the concepts and practical aspects of differential privacy for the data mining research community.

- We avoid detailed theoretical analysis of related differentially private algorithms and instead place more focus on its practical aspects which may benefit applications in the real world.
- We try to avoid repeating many references that have already been analyzed extensively in the above well-cited surveys.
- Even though differential privacy covers multiple research directions, we restrict our observations to data publishing and data analysis scenarios, which are the most popular scenarios in the data mining community.

Table 1.1 defines the scope of these two major research directions in differential privacy. Mechanism design for data publishing is normally independent from its publishing targets, as the goals of publishing is to release query answers or a dataset for further usage and, hence, is unknown to the curator. The mechanism design for data analysis aims to preserve privacy during the analysis process. The curator already knows the details of the analysis algorithm, so the mechanism is associated with the analysis algorithm.

Table 1.1 Comparison between differentially private data publishing and analysis

	Differentially private data publishing	Differentially private data analysis
Mechanism	Independent mechanism	Coupled with a particular algorithm
Input	Various data types	Transaction dataset (training samples)
Output	Query answers or datasets	Various models

Chapter 2
Preliminary of Differential Privacy

2.1 Notations

Figure 2.1 shows a basic attack model. Suppose a curator would like to preserve privacy for n records in a dataset D. However, an attacker has all information about $n-1$ records in dataset D except the nth record. These $n-1$ records can be defined as background information. He/she can make a query on the dataset D to get aggregate information about n records in D. After compare the difference between query result with the background information, the attacker can easily identify the information of record n.

Differential privacy aims to resist the attack. Differential privacy acquires the intuition that releasing an aggregated result should not reveal too much information about any individual record in the dataset. Figure 2.2 shows how differential privacy resists the attack. We define a dataset D' that differs with D with only one record, say, x_n. When the attacker make the query f on both datasets, he/she has a very high probability to get the same result s. Based on the results, he/she cannot identify whether x_n is in D or not. When the attacker cannot tell the difference between the

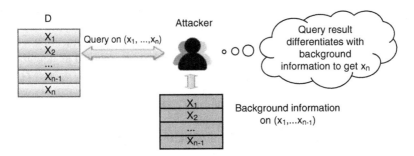

Fig. 2.1 Attacker model

© Springer International Publishing AG 2017
T. Zhu et al., *Differential Privacy and Applications*,
Advances in Information Security 69, DOI 10.1007/978-3-319-62004-6_2

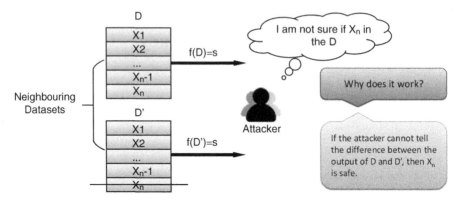

Fig. 2.2 Differential privacy

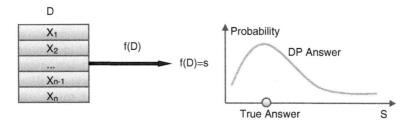

Fig. 2.3 Query answer distribution

outputs of D and D', then x_n is safe. If the property is applicable for all records in D, the dataset D can preserve privacy for all records.

In differential privacy, curator will not publish a dataset directly, instead, public users submit their statistical queries to the curator, and curator replies them with query answers. For a particular query, its true answer is unique, but its differentially private answer is a distribution, as shown in Fig. 2.3, dataset D and D' have very high probability to output same results.

To present the definition of differential privacy formally, we use the following key notations in the book. We consider a finite *data universe* $\mathscr{X}$ with the size $|\mathscr{X}|$. Let r represent a record with d attributes, a dataset D is an unordered set of n records sampled from the universe $\mathscr{X}$. Two datasets D and D' are defined as neighboring datasets if differing in one record. A query f is a function that maps dataset D to an abstract range $\mathbb{R}$: $f : D \rightarrow \mathbb{R}$. A group of queries is denoted as F. Normally, we use symbol m to denote the number of queries in F. There are various types of queries, such as count, sum, mean and range queries.

The target of differential privacy is to mask the difference of query f between the neighboring datasets [64]. The maximal difference on the results of query f is defined as the *sensitivity* Δf, which determines how much perturbation is required for the private preserving answer. To achieve the perturbation target, differential privacy provides a mechanism M accesses the dataset and implements

Table 2.1 Notations

Notations	Explanation	Notations	Explanation
$\mathcal{X}$	Universe	D	Dataset; training sample set
D'	Neighboring dataset	$\mathcal{D}$	Dataset distribution
r, x	Record in dataset; training sample	d	Dataset dimension
n	The size of dataset	N	The size of a histogram
f	Query	F	Query set
m	The number of queries in F	M	Mechanism
$\widehat{f}$	Noisy output	k	Represent some small value of constant
ϵ	Privacy budget	Δf	Sensitivity
G	Graph data	t, T	Time, time sequence, or iterative round
$\mathbf{w}$	Output model, or weight	$VC(\cdot)$	VC dimension
$\ell(\cdot)$	Loss function	α, β, δ	Accuracy parameter

some functionality. The perturbed output is denoted by a 'hat' over the notation. For example, $\widehat{f}(D)$ denotes the randomized answer of querying f on D. Table 2.1 summarizes some major notations used in the book. There are some other Greek symbols such as θ, η will be used temporarily in different chapters.

2.2 Differential Privacy Definition

A formal definition of differential privacy is shown below:

Definition 2.1 ((ϵ, δ)-Differential Privacy [67]) A randomized mechanism M gives (ϵ, δ)-differential privacy for every set of outputs S, and for any neighbouring datasets of D and D', if M satisfies:

$$Pr[M(D) \in S] \leq \exp(\epsilon) \cdot Pr[M(D') \in S] + \delta. \qquad (2.1)$$

Figure 2.4 shows mechanism on neighbouring datasets. For a particular output, the ratio on two probabilities is bounded by e^ϵ. If $\delta = 0$, the randomized mechanism M gives ϵ-differential privacy by its strictest definition. (ϵ, δ)-differential privacy provides freedom to violate strict ϵ-differential privacy for some low probability events. ϵ-differential privacy is usually called *pure differential privacy*, while (ϵ, δ)-differential privacy with $\delta > 0$ is called *approximate differential privacy* [20].

2.2.1 The Privacy Budget

In Definition 2.1, parameter ϵ is defined as the *privacy budget* [89], which controls the privacy guarantee level of mechanism M. A smaller ϵ represents a stronger privacy. In practice, ϵ is usually set as less than 1, such as 0.1 or ln 2. Two privacy

Fig. 2.4 Query answer
distribution

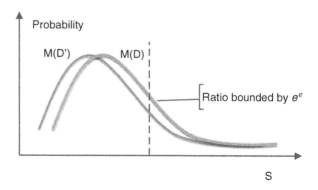

composition theorems are widely used in the design of mechanisms: sequential composition [157] and parallel composition [155], as defined in Theorem 2.1 and Theorem 2.2, respectively.

Theorem 2.1 (Parallel Composition) *Suppose we have a set of privacy mechanisms $M = \{M_1, \ldots M_m\}$, if each M_i provides ϵ_i privacy guarantee on a disjointed subset of the entire dataset, M will provide $(\max\{\epsilon_1, \ldots, \epsilon_m\})$-differential privacy.*
The *parallel composition* corresponds to a case where each M_i is applied on disjointed subsets of the dataset. The ultimate privacy guarantee only depends on the largest *privacy budget* allocated to M_i.

Theorem 2.2 (Sequential Composition) *Suppose a set of privacy mechanisms $M = \{M_1, \ldots M_m\}$ are sequentially performed on a dataset, and each M_i provides ϵ_i privacy guarantee, M will provide $(\sum_{i=1}^{m} \epsilon)$-differential privacy.*
The sequential composition undertakes the privacy guarantee for a sequence of differentially private computations. When a set of randomized mechanisms has been performed sequentially on a dataset, the final privacy guarantee is determined by the summation of total privacy budgets.

These two composition theorems bound the degradation of privacy when composing several differentially private mechanisms. Figure 2.5 shows their differences. Based on them, Kairouz [112] and Murtagh [164] provided optimal bounds when m mechanisms are adaptive, which means that M_i will be designed based on the result of M_{i-1}. They claimed that the privacy budgets will be consumed less when the mechanisms are adaptive. Currently, however, the parallel and sequential composition are most prevalent and straightforward way to analysis the privacy budget consuming of a privacy preserving algorithm.

2.3 The Sensitivity

Sensitivity determines how much perturbation is required in the mechanism. For example, when we publish a specified query f of dataset D, the sensitivity will calibrate the volume of noise for $f(D)$. Two types of sensitivity are employed in differential privacy: the *global sensitivity* and the *local sensitivity*.

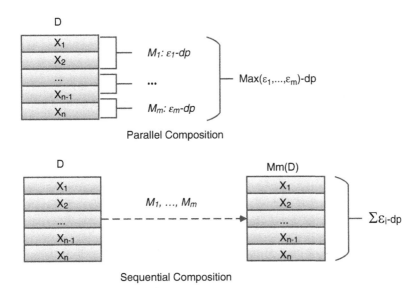

Fig. 2.5 Privacy budget composition

2.3.1   The Global Sensitivity

The global sensitivity is only related to the type of query f. It considers the maximum difference between query results on neighboring datasets. The formal definition is as below:

Definition 2.2 (Global Sensitivity)   For a query $f : D \to \mathbb{R}$, the *global sensitivity* of f is defined as

$$\triangle f_{GS} = \max_{D,D'} ||f(D) - f(D')||_1. \tag{2.2}$$

Global sensitivity works well when queries have relative lower sensitivity values, such as count or sum queries. For example, the count query normally has $\triangle f_{GS} = 1$. When the true answer is over hundreds or thousands, the sensitivity is much lower than the true answer. But for queries such as median, average, the global sensitivity yields high values comparing with true answers. We will then resort to local sensitivity for those queries [172].

2.3.2   The Local Sensitivity

Local sensitivity calibrates the record-based difference between query results on neighboring datasets [172]. Comparing with the global sensitivity, it takes both the record and query into consideration. The local sensitivity is defined as below:

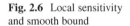
Fig. 2.6 Local sensitivity and smooth bound

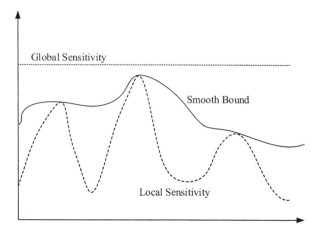

Definition 2.3 (Local Sensitivity) For query $f : D \rightarrow \mathbb{R}$, *local sensitivity* is defined as

$$\triangle f_{LS} = \max_{D'} ||f(D) - f(D')||_1. \tag{2.3}$$

For many queries, such as the `median`, the local sensitivity is much smaller than the global sensitivity. However, as the changing of local sensitivity may result in information disclosure, it cannot be used in mechanisms directly. The value of local sensitivity should be changed smoothly, so that a smooth bound should be added.

Definition 2.4 (Smooth Bound) For $\beta > 0$, a function $B : D \rightarrow \mathbb{R}$ is a β-smooth upper bound on the local sensitivity of f if it satisfies the following requirements,

$$\forall D \in X : B(D) \geq f_{LS}(D) \tag{2.4}$$

$$\forall D, D' \in X : B(D) \leq e^{\beta} S(D). \tag{2.5}$$

Figure 2.6 shows the relationship between the local sensitivity, smooth bound and the global sensitivity. For some queries, the local sensitivity is lower than global sensitivity. For queries such as `count` or `range`, the local sensitivity is identical to global sensitivity. Because most literatures were concerned with the global sensitivity, without specification, sensitivity refers to global sensitivity in this book.

2.4 The Principle Differential Privacy Mechanisms

Any mechanism meeting Definition 2.1 can be considered as differentially private. Currently, three basic mechanisms are widely used to guarantee differential privacy: the Laplace mechanism [68], the Gaussian mechanism [72] and the exponential mechanism [157]. The Laplace and Gaussian mechanisms are suitable for numeric queries and the exponential mechanism is suitable for non-numeric queries.

2.4.1   The Laplace Mechanism

The Laplace mechanism relies on adding controlled Laplace noise to the query result before returning it to the user. The noise is sampled from the Laplace distribution, which is centered at 0 with scaling b. The noise is presented by $Lap(b)$, in which a larger b indicates a higher noise. The corresponding probability density function is:

$$Lap(x) = \frac{1}{2b} \exp(-\frac{|x|}{b}). \qquad (2.6)$$

The mechanism is defined as follows:

Definition 2.5 (Laplace Mechanism)   Given a function $f : D \rightarrow \mathbb{R}$ over a dataset D, mechanism M provides the ϵ-differential privacy if it follows Eq. (2.5)

$$M(D) = f(D) + Lap(\frac{\Delta f}{\epsilon}). \qquad (2.7)$$

The mechanism shows that the size of noise is related to the sensitivity of query f and the privacy budget ϵ. A larger sensitivity leads to a higher volume of noise.

2.4.1.1   The Gaussian Mechanism

To achieve (ϵ, δ)-differential privacy, one can use Gaussian noise [72]. In this case, rather than scaling the noise to the ℓ_1 sensitivity, one instead scales to the ℓ_2 sensitivity as follow Definition 2.6:

Definition 2.6 (ℓ_2-Sensitivity)   For a query $f : D \rightarrow \mathbb{R}$, the ℓ_2-sensitivity of f is defined as

$$\Delta_2 f = \max_{D, D'} ||f(D) - f(D')||_2. \qquad (2.8)$$

The Gaussian mechanism with parameter σ adds zero-mean Gaussian noise with variance σ.

Definition 2.7 (Gaussian Mechanism)   Given a function $f : D \rightarrow \mathbb{R}$ over a dataset D, if $\sigma = \Delta_2 f \sqrt{2 ln(2/\delta)}/\epsilon$ and $\mathcal{N}(0, \sigma^2)$ are i.i.d. Gaussian random variable, mechanism M provides the ϵ, δ-differential privacy when it follows Eq. (2.7)

$$M(D) = f(D) + \mathcal{N}(0, \sigma^2). \qquad (2.9)$$

The Gaussian mechanism follows the same privacy composition to the Laplace mechanism.

2.4.2 The Exponential Mechanism

For non-numeric queries, differential privacy uses an exponential mechanism to randomize the results, and this is paired with a score function $q(D, \phi)$ that represents how good an output ϕ is for dataset D. The choice of score function is application dependent and different applications lead to various score functions. The Exponential mechanism is formally defined as below:

Definition 2.8 (Exponential Mechanism) Let $q(D, \phi)$ be a score function of dataset D that measures the quality of output $\phi \in \Phi$. Then an Exponential mechanism M is ϵ-differential privacy if

$$M(D) = \{\text{return } \phi \text{ with the probability} \propto \exp(\frac{\epsilon q(D, \phi)}{2\Delta q})\}. \tag{2.10}$$

where Δq represents the *sensitivity* of score function q.

2.4.2.1 Mechanism Example

An example is presented below to illustrate some fundamental concepts of the sensitivity, privacy budget and mechanisms. Suppose Table 2.2 shows a medical dataset D of a district, and differential privacy mechanism M will guarantee the privacy of each individual in D.

Suppose query f_1 asks: *how many people in this table have HIV?* Because the query result is numeric, we can use Laplace mechanism to guarantee differential privacy. First, we analyse the sensitivity of f_1. According to Definition 2.2, deleting a record in this D will affect the query result maximally by 1. The sensitivity of f_1 is $\Delta f_1 = 1$. Second, we choose a privacy budget ϵ for the Laplace mechanism. Suppose we set $\epsilon = 1.0$. According to Definition 2.5, the noise that sample from

Table 2.2 Medical record

Name	Job	Gender	Age	Disease
Alen	Engineer	Male	25	Flu
Bob	Engineer	Male	29	HIV
Cathy	Lawyer	Female	35	Hepatitis
David	Writer	Male	41	HIV
Emily	Writer	Female	56	Diabetes
...	...	...	...	...
Emma	Dancer	Female	21	Flu

Table 2.3 Medical record exponential mechanism output

Options	Number of people	$\epsilon = 0$	$\epsilon = 0.1$	$\epsilon = 1$
Diabetes	24	0.25	0.32	0.12
Hepatitis	8	0.25	0.15	4×10^{-5}
Flu	28	0.25	0.40	0.88
HIV	5	0.25	0.13	8.9×10^{-6}

$Lap(1)$ will be added to the true answer $f_1(D)$. Lastly, the mechanism M will output a noisy answer $M(D) = f_1(D) + Lap(1)$. If the true answer is 10, the noisy answer might be 11.

Suppose we have another query f_2: *what is the most common disease in this district?* This query will generate non-numeric result and we can apply the exponential mechanism. Table 2.3 lists all the diseases and their true numbers in the first two columns. We first define the score function of f_2. We adopt the number of people on each disease as the score function q. As deleting a person will have a maximum impact of 1 on the result of q, the sensitivity of q is $\triangle q = 1$. The probability of the output can then be calculated by Definition 2.8. Table 2.3 lists the results when $\epsilon = 0$, $\epsilon = 0.1$ and $\epsilon = 1$.

In the third column of Table 2.3, $\epsilon = 0$ means that the mechanism chooses an answer uniformly from those four options. The output probabilities are equal in these options. Obviously, $\epsilon = 0$ provides the highest privacy level, however it loses almost all the utility. When $\epsilon = 0.1$, *Flu* has the highest probability of being chosen and *HIV* has the lowest probability. The gap is not very large, which indicates that it can provide acceptable privacy and utility levels. When $\epsilon = 1$, the probability gap between *HIV* and other diseases is significant, which means that the mechanism can retain a high utility, but have a lower privacy level.

2.5 Utility Measurement of Differential Privacy

When privacy level is fixed to ϵ, several utility measurements have been used in both data publishing and analysis to evaluate the performance of differential privacy mechanisms.

- *Noise size measurement*: the easiest way is calibrating how much noise has been added to the query results. A smaller noise indicates higher utility. This utility measurement has been widely used in data publishing.
- *Error measurement*: utility can be evaluated by the difference between the non-private output and the private output. For example, the utility of single query publishing can be measured by $|f(D) - \widehat{f}(D)|$. A smaller distance shows higher utility. The error measurement is normally represented by a bound with accuracy parameters [29]:

Definition 2.9 ((α,β)-Useful)   A set of query F is (α,β)-utility if

$$Pr(\max_{f\in F}|F(D) - \widehat{F}(D)| \leq \alpha) > 1 - \beta, \tag{2.11}$$

where α is the accuracy parameter that bounds the error.

For different publishing scenarios, the error measurement can be interpreted in various ways. For synthetic dataset publishing, Eq. (2.11) can be interpreted to:

$$Pr(\max_{f\in F}|F(D) - F(\widehat{D})| \leq \alpha) > 1 - \beta. \tag{2.12}$$

For data analysis, the utility normally depends on the analysis algorithms. Suppose the algorithm is denoted by M and the private algorithm is denoted by $\widehat{M}$, Eq. (2.11) can be interpreted to

$$Pr(|M(D) - \widehat{M}(D)| \leq \alpha) > 1 - \beta. \tag{2.13}$$

Equation (2.13) has several implementations in data analysis, such as risk bound and sample complexity.

Chapter 3
Differentially Private Data Publishing: Settings and Mechanisms

3.1 Interactive and Non-interactive Settings

Differentially private data publishing aims to output aggregate information to the public without disclosing any individual's record. This problem can be presented as follows: if a curator has a dataset D and receives a query set $F = \{f_1, \ldots, f_m\}$, he/she is required to answer each query $f_i \in F$ subject to the constraints of differential privacy. Two settings, interactive and non-interactive, are involved in this publishing scenario: in the interactive setting, a query f_i can not be issued until the answer to the previous query f_{i-1} has been published. In the non-interactive setting, all queries are given to the curator at one time. The curator can provide answers with full knowledge of the query set. Figure 3.1 shows both interactive and non-interactive settings.

An example is presented in Tables 3.1 and 3.2 to show the difference between the two settings. Suppose a curator has a medical dataset D, queries to the curator may presented as follows:

- f_1: *How many patients have diabetes at the age of* 40–79?
- f_2: *How many patients have diabetes at the age of* 40–59?

Suppose the privacy budget ϵ is fixed for each query. In the interactive setting, the curator will first get f_1, then counts the number of patients who have diabetes with the age from 40 to 79 and adds an independent Laplace noise with sensitivity equal to 1, $Lap(1/\epsilon)$, to the number. When f_2 is then submitted to the curator, f_2 will be answered with sensitivity equal to 2, as changing one person in the table will maximize change results of both queries. The total noise added to the query set is $Lap(1/\epsilon) + Lap(2/\epsilon)$.

In the non-interactive setting, both queries will be submitted to the curator at the same time. The sensitivity measured for both queries is 2. The total noise added to the query set is $2 * Lap(2/\epsilon)$, which is larger than the interactive setting.

© Springer International Publishing AG 2017
T. Zhu et al., *Differential Privacy and Applications*,
Advances in Information Security 69, DOI 10.1007/978-3-319-62004-6_3

Fig. 3.1 Interactive and
non-interactive settings

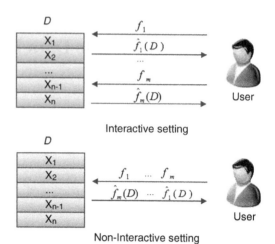

Table 3.1 Medical table

Name	Age	Diabetes
Alen	25	N
Bob	29	N
Cathy	35	Y
David	41	Y
Emily	56	N
…	…	…
Emma	21	Y

Table 3.2 Frequent table

Age	Diabetes number	Variable
60–79	41	x_1
40–59	32	x_2
20–39	8	x_3
0–19	1	x_4

The correlation between queries leads to a higher sensitivity. Therefore, the non-interactive setting normally incurs more noise than the interactive setting.

The above example presents the difference between two settings, and shows the size of noise increasing dramatically when queries are correlated to each other. In addition, for a dataset with size n, the Laplace mechanism can only answer, at most, sub-linear in n number of queries to a certain level of accuracy [59]. To simplify the problem, most papers on interactive setting assume that those queries are independent to each other. These weaknesses make the Laplace mechanism impractical in the scenarios that require answering large amounts of queries. New mechanisms are required.

To fix the weaknesses of the Laplace mechanism, researchers are concerned with new mechanisms design, aiming to publish various types of data with limited noise. Table 3.3 summarizes the problem characteristics of differentiallyprivate data

Table 3.3 Differentially private data publishing problem characteristics

Differentially private data publishing	
The nature of input data	Transaction, histogram, graph, stream
The nature of output data	Query result, synthetic dataset
Publishing setting	Interactive, non-interactive
Publishing mechanism	Laplace/exponential, query separation, transformation, iteration, partition of dataset
Challenges	Query number, accuracy, computational efficiency

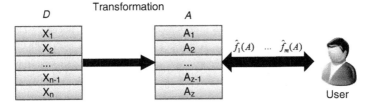

Fig. 3.2 Transformation

publishing, in which mechanisms design is focusing on the number of queries, the accuracy of the output, and the computational efficiency. Among them, the number of queries means the maximum number of queries the mechanism can answer, indicating the capacity of the mechanism. The accuracy of the publishing output is measured by the expected error between the randomized answer and the true answer. Furthermore, if the running time of answering a query is polynomial in n and d, the publishing mechanism is considered to be efficient.

3.2 Publishing Mechanism

We categorize the existing mechanisms into several types: transformation, partitioning of dataset, query separation and iteration. Tables 3.1 and 3.2 are used again to show the key idea of those types.

- *Transformation*: The transformation mechanism maps the original dataset to a new structure to adjust the sensitivity or noise. The key issue in the transformation mechanism is to find a suitable structure that minimizes the error of query answers. Figure 3.2 shows the basic process.

  In the above example, the original dataset can be transferred to a frequent dataset as shown in Fig. 3.2.

Table 3.4 Differentially private data publishing mechanism comparison

Mechanism	Description	Advantage	Challenge
Laplace	Add Laplace noise directly to the query	Easy to implement; can answer all types of real-value queries	Answer a sub-linear in n number of queries; introduce large volume of noise
Transformation	Transfers original dataset to a new structure. The sensitivity of the query set will be adjusted	Noise can be reduced; consistency of results can be maintained	Not easy to find a new structure that suits for queries
Dataset partitioning	Divides dataset into several parts and adds noise to each part separately	Sensitivity can be decreased which results in less error in the output	Partition strategy is not easy when answering multiple queries
Iteration	Updates datasets or query answers recursively to approximate the noisy answer	Only some updates will consume privacy budget, so more queries can be answered (linear to n or exponential number of n) in a fixed privacy budget	Most iteration mechanisms are computationally inefficient; unsuitable parameters can result in inferior performance
Query separation	Only need to add noise to small numbers of queries	Some queries do not consume privacy budget, so the mechanism can answer more queries with a fixed privacy budget	Separating queries is a difficult problem

In the new structure, f_2 can be answered by the second row with the sensitivity equals to 1. And we can get the result of diabetes from age 60 to 79 independently from the first row in Table 3.2. As both results are independent to each other, the result of f_1 can be answered by the result of $\widehat{f_2}$ and the first row. The total noise of two queries will be $3 * Lap(1/\epsilon)$, which is lower than the non-interactive Laplace mechanism. In this example, the new structure is used to decompose the correlation between queries, so the sensitivity can be decreased as well. The challenge is to find a new structure. In this example, the new structure is a meaningful frequent table; however, in most cases, the new structure is meaningless, only used to decrease the sensitivity.

- *Partition of dataset*: the original dataset is divided into several parts to decrease noise. In the above example, suppose we need to answer f_1 with Table 3.2, the noise $Lap(1/\epsilon)$ needs to be added twice: one is added to the first row, another is added to the second row. The total added noise will be $2 * Lap(1/\epsilon)$. If we partition the dataset by another way, for example, arranging the age range to 40–79, the total noise will be decreased to $Lap(1/\epsilon)$. The challenge is how to design the partition strategy with multiple queries.

- *Query separation*: query separation assumes that a query set can be separated into several groups, and that some queries can be answered in the sense of reusing noise. In the above example, if f_2 has been answered, f_1 can be approximately answered by doubling the answer of f_2 as the age range is doubled. Query separation is a strategy to break limitations on the number of queries.

- *Iteration*: iteration is a mechanism that updates a dataset recursively to approximate the noisy answers for a set of queries. For example, we can manually define an initial dataset D_0, in which the number of diabetes in Table 3.2 at different age ranges are equal, and then perform f_1 on D_0 and compare the noise result $\widehat{f}_1(D)$ with $f_1(D_0)$. If the distance between the two answers is smaller than a pre-defined threshold, $f_1(D_0)$ can be published, and D_0 will be used in the next round. Otherwise, $\widehat{f}_1(D)$ will be published and D_0 will be updated by a particular strategy into D_1. As publishing $f_1(D_0)$ does not consume any privacy budget, the iteration mechanism can achieve a higher utility and can answer more queries than the Laplace mechanism. The challenges are how to design the update strategy and how to set related parameters such as threshold.

Table 3.4 compares the various publishing mechanisms. In following subsections, we present how those mechanisms work in both interactive and non-interactive settings.

Chapter 4
Differentially Private Data Publishing: Interactive Setting

Interactive settings operate on various aspects of the input data, including transactions, histograms, streams and graph datasets. In the following sections, we discuss publishing scenarios involving these types of input data.

4.1 Transaction Data Publishing

The most popular representation on D is the transaction dataset, in which every record represents an individual with d attributes. Table 4.1 shows a typical transaction dataset with $d = 3$. Diabetes is considered to be sensitivity information that needs to be preserved. The transaction dataset is the most prevalent nature of data and attracts significant attention. Several mechanisms are proposed in this area, including Laplace, transformation, query separation and iteration.

4.1.1 Laplace

Dwork et al. [68] proposed the Laplace mechanism to publish the transaction dataset in their initial work. As mentioned earlier, this mechanism can efficiently answer all types of real-values queries, but the maximum number of queries is limited by the sub-linear size of the dataset. Dinur et al. [59] proved that answering substantially more than a linear number of subset sum queries with error $o(n^{1/2})$ yields *blatant non-privacy*. This lower bound implied that for a dataset with size n, the Laplace mechanism can only answer maximum sub-linear of n queries in a certain level of accuracy. Otherwise, adversaries can reconstruct a $1 - o(1)$ fraction of the original database.

© Springer International Publishing AG 2017   23
T. Zhu et al., *Differential Privacy and Applications*,
Advances in Information Security 69, DOI 10.1007/978-3-319-62004-6_4

Table 4.1 Transaction dataset

Name	Age	Has diabetes?
Alice	35	1
Bob	50	1
Cathy	12	0
...	...	...
Eva	35	0

The accuracy of the Laplace mechanism output can easily be estimated by analyzing the property of Laplace distribution. For every sequence of m queries with sensitivity Δf, the accuracy is bounded by $O(\frac{\Delta fm \log m}{\epsilon})$.

Because of the weaknesses of the query number limitation and inferior accuracy on large sets of queries, the Laplace mechanism is now regarded as a baseline mechanism in transaction data release and new mechanisms need to be proposed.

4.1.2 Transformation

Transformation changes or projects the current transaction data to another structure on the assumption that sensitivity can be diminished in the new structure. Hardt et al. [93] proposed the k-norm method, which considers a dataset to be a unit ball and linear queries to be a set of linear mappings. The query result is transformed into a convex polytope and the exponential mechanism is used by k-norm to sample points in this convex polytope as query results. K-norm measures the bounded error to $O(\min\{\frac{m}{\epsilon}, \sqrt{m \log(\frac{n}{m})}/\epsilon\})$ per answer, which is an improvement over the Laplace mechanism. However, because k-norm uses an exponential mechanism and samples random points from high-dimensional convex bodies, the computation time is inefficient.

4.1.3 Query Separation

The goal of query separation is to design a separation strategy for given types of queries to decrease noise. Roth [195] presented the *median* mechanism and found that, among any set of m queries, there are $O(\log m \log |\mathcal{X}|)$ queries that can determine the answers of all other queries. Based on this observation, all queries are separated into hard and easy queries. Hard queries can be answered directly by the Laplace mechanism, while easy queries are answered by the median values of hard query results. Therefore, easy queries do not consume any privacy budget. By separating the queries, the median mechanism can answer exponentially many more queries with acceptable accuracy; however, it is inefficient and comes with an exponential time complexity corresponding to the dataset size n.

Muthukrishnan et al. [165] particularly considered range queries and decomposed the range space into logarithmic number of smaller distinct ranges. They exploited the balance between the number of distinct ranges and the maximum range size using discrepancy theory. Ultimately, they improved the mean squared error of range queries to $O(\log n)^{1-O(1)}$.

4.1.4 Iteration

Hardt et al. [92] proposed *private multiplicative weights* (PMW), which considers datasets as a histogram with positive weight on each bin. By updating the weights, PMW constructs a histogram sequence to answer a set of queries. Specifically, the initial histogram x_0 was set as a uniform distribution over the domain. The mechanism then maintained a sequence of histogram $x_0, x_1, \ldots, x_t$ in t iterations, which gave increasing approximation to the original histogram x. After the parameters have been calibrated for complexity and accuracy, this mechanism is able to answer each query with a sampling error approximately to $O((\log m)/\sqrt{n})$. This means that the sampling error grows logarithmically with an increase in the number of queries being answered, while the Laplace mechanism's error is linear, increasing by m. In addition, PMW can accurately answer an exponential number of queries.

Similarly, Gupta et al. [88] presented a general iteration framework termed iterative database construct (IDC), which implements other release mechanisms by using the framework. In each round of iteration, when a significant difference between the current dataset and the original dataset is witnessed for a given query, the mechanism updates the current dataset for the next update. The update function was defined by the Frieze/Kannan low-rank matrix decomposition algorithm. The effectiveness of the framework is evaluated by cut queries in a social network graph dataset. IDC is a more general framework that can be incorporated into various other mechanisms, including the PMW and median.

4.1.5 Discussion

The transformation and query separation mechanisms can answer more queries than the Laplace mechanism in a fixed privacy budget. They however have some very restrict criteria on the dataset or queries. For example, the transformation mechanism requires the dataset has some special properties and the query separation mechanism assumes the query can be divided into distinct types. These criteria make these mechanisms impractical.

The iteration mechanism has little assumption on the dataset and has several advantages over the various publishing mechanisms mentioned above, such as decreasing noise when confronted with many queries; it has a lower running time in practice for low-dimensional datasets; it can be easily implemented and perform parallel on datasets. Many subsequent works therefore followed this mechanism. For example, Ullman [222] extended PMW to convex minimization to exponentially answer many *convex minimization* queries. Based on IDC, Huang et al. [103] presented an efficient query on distance defined over an arbitrary metric. However, since the iteration mechanism utilizes histogram to represent the dataset, it can not be applied on complex or high-dimensional datasets.

Even most mechanisms focused on count query, there has also been significant attention paid to the specific type of queries, such as conjunctions query, cut query, distance query [103], range query, and halfspace queries. It shows that excluding the Laplace mechanism, the other mechanisms can answer large numbers of queries (exponential to n) with acceptable accuracy.

However, there is a serious problem in computational efficiency for those mechanisms. Most works have an exponential running time, which implies that the computational efficiency needs to be sacrificed for acceptable accuracy. This conclusion was confirmed by Dwork et al. [66], who claimed that computationally inefficient mechanism can accurately answer an exponential number of adaptively chosen statistical queries. Ullman [221] quantitatively analyzed the bound and showed that a differential privacy algorithm will require exponential running time to answer $n^{2+o(1)}$ statistical queries. Hardt et al. [94] improved the bound and showed that there is no computationally efficient algorithm that can give valid answers to $n^{3+o(1)}$ (cube) adaptively chosen statistical queries from an unknown distribution.

4.2   Histogram Publishing

It is often convenient to regard transaction data in terms of their histogram representations. Suppose a histogram has N bins, a differential privacy mechanism aims to hide the frequency of each bin. The advantage of histogram representation is that limits the sensitivity to noise [62]. For example, when the histogram serves to support the range or count queries, adding or removing a single record will affect, at most, one bin. Hence, range or count queries on the histogram have a sensitivity equal to 1, and the volume of added noise to each bin will be relatively small.

A transaction dataset can be mapped to a histogram over the domain $\mathscr{X}$. Each bin represents the combination of single or several attributes. The frequency of each bin is the fraction of its count in the original dataset.

Formally, we define the histogram representation as below:

Definition 4.1 (Histogram Representation)   A dataset D can be represented by a histogram x in a domain $\mathscr{X}$: $x \in N^{|\mathscr{X}|}$, where N consists of all possible combinations of attributes.

4.2.1 Laplace

This is a direct mechanism which adds Laplace noise to the frequency of each bin that a query covered. When a count of range query covers only small number of bins, this mechanism retains high utility for the query result; however, if the original dataset contains multiple attributes, the combination of these attributes and their related range of values will lead to a large number of bins. The answer of queries are meaningless due to the large amount of error accumulated.

4.2.2 Partition of Dataset

The most popular mechanism for histogram release is the partition of the dataset, except for the traditional Laplace mechanism. As the number of bins is derived from the partition of attribute values, one method for decreasing error is to optimize the partition strategy. Figure 4.1 shows the histogram example. Figure 4.1a is an histogram example. When publish the histogram in the constraint of differential privacy, as shown in Fig. 4.1b, the noise is estimated as $\eta = Lap(1/\epsilon)$, which has to be added to each bin. Multiple bins lead to large noise. Figure 4.1c illustrates that if bins can be re-partitioned, for example, range of age can be merged from [0–20], [20–40] to [0–40], noise will be diminished. After that, original frequency of bin [0–20] can be estimated by dividing the frequency of bin [0–40], as shown in Fig. 4.1d. However, splitting the large bin into smaller bins leads to more Laplace noise, while estimating the proportion of the large bin's frequency may introduce estimation error. Therefore, optimizing the partition strategy to obtain a trade-off between splitting and merging bins is a challenge that needs to be addressed.

Xiao et al. [240] provided a *kd-tree* based partition method to generate nearly uniform partitions to decrease the average error. Their idea is to partition the original histogram and merge bins with similar frequencies. The average frequency will then be close to those original frequencies, and will reduce the average error. Xiao et al. apply the *kd-tree* to identify bins which have similar frequencies when answering queries. A *kd-tree* is a binary tree that every non-leaf node can be considered as a splitting hyperplane to divide the space into two half-spaces. Generating a balanced *kd-tree* on the histogram frequency can help to divide the original histogram into a small number of almost uniform structures. The authors achieved more accurate result than Laplace mechanism.

In their subsequent work [239], Xiao et al. extended the two-dimensional histogram into a d-dimensional ($d > 2$) one by using an algorithm *DPCube*. The authors implemented a set of query matrices to generate an optimal query strategy on a d-dimensional histogram to test performance. When the parameters (frequency closeness threshold, partition times) are estimated properly, the *DPCube* achieves a good balance between the maximum number of queries and the introduced errors.

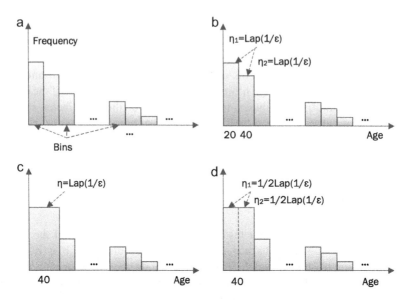

Fig. 4.1 Histogram publishing. (**a**) Histogram. (**b**) Noisy histogram. (**c**) Bin partition. (**d**) Estimate frequency

Xu [243] provided two partition strategies by minimizing the sum of squared error (SSE) of a set of queries. Both strategies set each attribute as a single bin at an initial state and partition the attribute value to create more bins. The first strategy, *NoiseFirst*, injects Laplace noise into each bin before partitioning the attribute values. Another strategy, *StructureFirst*, uses an exponential mechanism to select optimal splitting values of attributes by adopting the SSE as the score function.

Qardaji et al. [185] partitioned the attribute values in a hierarchical way. They also focus on range queries and allocate ranges into a tree. The root of the tree is the full range of values of an attribute or several attributes. Each node in the tree is associated with the union of the ranges of its children. Several unit-length ranges are defined as leaves. On each branch of the tree, a factor controls the accuracy of the query result. These factors are further studied with a mean squared error (MSE) when answering range queries, and the results are optimized by tuning these factors.

4.2.3 Consistency of Histogram

When noise is added to bins, the consistency of the query results may be destroyed. For example, the sum of two bins $A + B$ may less than the frequency of one bin A. To maintain consistency of the histogram release, Hay et al. [96] defined a *constrained inference* to adjust the released output. Two types of consistency constraints were explored. The first, *sorted constraints*, requires query results to satisfy a particular sequence. The second, *hierarchical constraints*, predefines the sequence of a hierarchical interval set. Their proposed approach provides a noisy answer set to respond to the query set by using a standard differential privacy mechanism. The *constrained inference* step applies the linear combination method to estimate a set of approximate answers that are close to noisy answers, which satisfy the consistency constraints. An approximate answer for the query set F is ultimately released, which can preserve the differential privacy and retain the consistency between results.

There are some other ways to improve the consistency of histogram. For example, Lin et al. [146] applied a Bayesian approach to estimate sorted histograms and pointed out that substantial improvements in accuracy can be obtained if the estimation procedure makes use of knowledge on the noise distribution, proposed an estimation algorithm that views sorted histograms as a Markov chain and imposes ordering constraints on the estimates. Lee et al. [135] added a post-processing step before the histogram releasing. The post-processing step is formulated as a constrained maximum likelihood estimation problem, which is equivalent to constrained L_1 minimization.

4.3 Stream Data Publishing

In a real world scenario, it is practical for release information to be continuously updated as new data arriving. For example, the search log, recommender system, or location datasets increase continuously over time. These types of data can be modeled as stream data, which can be simplified as a bit string $\{0, 1\}^n$ and each 1 in the stream represents the occurrence of an event [70]. The bit stream normally associates with the continual release scenario, in which a differentially private mechanism releases one bit at every time step.

Figure 4.2 illustrates this continual release scenario. Suppose there is a binary bit stream $D \in \{0, 1\}^T$, where T represents a time sequence $T = \{t_k : k = 0, \ldots\}$. The bit $\sigma(t_k) \in \{0, 1\}$ denotes whether an event has occurred at time t_k, and $D(t_k)$ represents the prefix of a stream at time t_k. At each time step t_k, the number count of $1s$ is denoted as $\iota(t_k)$ and the mechanism outputs the noisy count is $\widehat{\iota}(t_k)$. Hence, the research issue is how to increase the accuracy of the released count under the constraint of differential privacy.

Fig. 4.2 Stream dataset

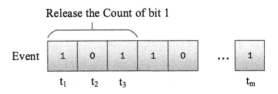

4.3.1 Laplace

Works on continual data publishing define two private levels in terms of differential privacy, namely *event-level* and *user-level* privacy [69]. The former hides a single event while the latter masks all events of a user. It appears that the *user-level* privacy leads to more error than the *event-level* privacy because a user may have several events to hide. In practice, most of the work focus on *event-level* privacy.

There are two ways the Laplace mechanism can be directly utilized for continual release. The first method divides the privacy budget into several pieces $\epsilon' = \epsilon/T$, and allocates them to each time step t. The mechanism samples fresh independent random noise $\zeta_t \sim Lap(1/\epsilon')$ at each time step, and releases a noise count by using $\widehat{\iota_t} = \iota(t) + \zeta_t$. Here, the total volume of noise is $T \cdot Lap(T/\epsilon)$, which is a dramatically large volume of noise that results in inferior utility.

The second method adds independent Laplace noise $\gamma_t \sim Lap(1/\epsilon)$ to each bit in the stream and at each time step t, computes $a_t = \sigma(t) + \zeta_t$. The final noisy count is $\widehat{\iota_t} = \sum_i^t a_i$. In a fixed total time step T, the introduced noise is $\sum_{i<T} \zeta_i$, which is bounded by $O(\frac{T \log(T/\delta)}{\epsilon})$. This result is better than that can be achieved by the first method, but the volume of noise is still too large.

4.3.2 Partition of Dataset

Chan et al. [34] presented a *p-sums* mechanism which computes the partial sum of consecutive bits in a stream. The results can be considered as intermediate results from which an observer can estimate the count at every time step. Laplace noise is added to a *p-sum* result rather than an individual count answer. This guarantees an error bound of $O(\frac{\log^{3/2} T}{\epsilon})$, which decreases the linear complexity of noise to *log* complexity.

4.3.3 Iteration

Dwork et al. [69] proposed a continual output transformation algorithm, which developed an iterative release mechanism to output count. The error is decreased to $O(\frac{\log^{3/2} T}{\epsilon})$.

Georgios et al. [123] argued that *event-level* privacy will disclose sensitive information while *user-level* privacy will destroy the utility of the data in the long run. They provided a notion of *w-event* privacy to achieve a balanced privacy target. They formulated the iteration mechanism by using a sliding window methodology with two privacy budget allocation schemes. This *w-event* privacy protects any event sequence from occurring within any window of *w* time steps.

4.3.4 Discussion

Even though continual data publishing is a popular topic in the data mining or machine learning community, there is still little research work that focuses on the privacy-preserving perspective. In fact, there are lots of unsolved problems in this area. For example, the issues of *how to release a multiple dimensional dataset periodically* and *how to deal with other statistical queries* in the continual release need further exploration.

4.4 Graph Data Publishing

With the significant growth of *Online Social Networks* (OSNs), the increasing volumes of data collected in those OSNs have become a rich source of insight into fundamental societal phenomena, such as epidemiology, information dissemination, marketing, etc. Most of OSN data are in the form of graphs, which represent information such as the relationships between individuals. We follow the convention to model an input OSNs dataset as a simple undirected graph $G = (V, E)$, where V is the set of nodes and $E \subseteq V \times V$ is the set of edges. The time of connection between nodes are defined as the *degree*. Figure 4.3 shows a graph.

Analyzing those graph data has enormous potential social benefits, but the fact that the same graph data in OSNs might be used to infer sensitive information about a particular individual [13] has raised concern among OSN participants. The analysis of OSNs is a rapidly growing field, and many models require the operation of *subgraph counting*, which counts the number of occurrences of a query subgraph in an input graph. For example, the subgraph counts of k-star and triangle, as well as the number of edges, are sufficient statistics for several popular exponential random graph models. Moreover, many descriptive statistics of graphs are functions of subgraph counts [113].

The privacy issue is more serious for graph data. For transaction dataset records, at least the attributes are given. While in graph data, the attributes are derived, not given. For example, given a graph, what are the important attributes to hide? The

Fig. 4.3 Graph data

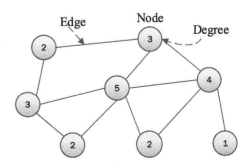

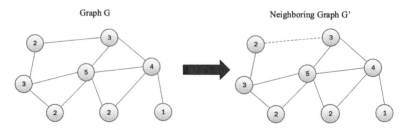

Fig. 4.4 Edge differential privacy

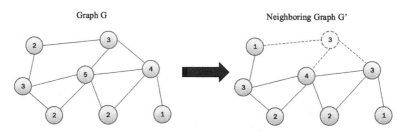

Fig. 4.5 Node differential privacy

node degree is an obvious attribute. But there can be many others, such as degree pair of a link, common neighbors between each pair of nodes, etc. These attributes are all derived. It is unclear what are the attributes the attackers can use to make the attack. It is extremely difficult to decide what graph attributes can be derived and need to be hidden.

In this case, two notions of neighboring graph can be defined in the context of differential privacy for graph data: edge neighboring and node neighboring. Accordingly, there are two concepts of differential privacy for graph: (*a*) *Edge differential privacy* means adding or deleting a single edge between two nodes in the graph makes negligible difference to the result of the query. Figure 4.4 shows the edge differential privacy. (*b*) *Node differential privacy* ensures the privacy of a query over two neighbouring graphs where two neighbouring graphs can differ up to all edges connected to one node. Figure 4.5 shows the node differential privacy. If the differential privacy mechanism is adopted in graph data, the research problem is then to design efficient algorithms to release statistics about graph while satisfying the definition of differential privacy.

4.4.1 Edge Differential Privacy

The first differential privacy research over graph data was conducted by Nissim et al., who showed how to estimate the number of triangles in a graph with edge differential privacy [172]. They introduced the idea of instance-dependent noise, and utilized the *smooth sensitivity*, which upper bounds the local sensitivity tightly, especially when the number of triangles varies smoothly in a neighborhood of the input graph. They showed how to efficiently calibrate noise in the context of subgraph counts, based on the *smooth sensitivity* of the number of triangles. The results of this technique were further extended by Karwa et al. [113] to release the counting on k-triangles and k-stars. They achieved ϵ-differential privacy for k-star counting, and (ϵ, τ)-differential privacy for k-triangle counting.

Rastogi et al. [190] studied the release of more general subgraph counts under edge adversarial privacy, They considered a Bayesian adversary whose prior knowledge is drawn from a specified family of distributions. By assuming that the presence of an edge does not make the presence of other edges, they computed a high probability upper bound on the local sensitivity, and then added noise proportional to that bound. Rastogi et al.'s method can release more general graph statistics, but their privacy guarantee protects only against a specific class of adversaries, and the magnitude of noise grows exponentially with the number of edges in the subgraph.

Hay et al. [95] considered publishing the degree distributions. They showed that the global sensitivity approach can still be useful when combined with post-processing of the released output to remove some added noise, and constructed an algorithm for releasing the degree distribution of a graph, with the edge differential privacy. They also proposed the notion of *node differential privacy* and highlighted the challenges in achieving it.

Zhang et al. [249] claimed that if one can find a isomorphic graph with proper statistical properties that similar to original graph, the isomorphic graph can be used to generate accurate query answers. Give a subgraph S, they adopted an exponential mechanism to search a number of isomorphic copies of S to answer subgraph queries.

4.4.2   Node Differential Privacy

Node differential privacy is a strictly stronger guarantee, but it is difficult to achieve because for many natural statistics, the sensitivities resulting from the change of one node are comparable to graph size.

With the observation that many useful statistics have low *global sensitivities* on graphs G_θ with small degree θ, the common approach to *node differential privacy* is to project the graph G on the degree-bounded graph G_υ, and evaluate the statistics on the G_υ [27, 117].

One straightforward method is to trunk the graph G by discarding nodes with degree $> \theta$. Given a query f defined on the trunked graph $\overline{G}$, Kasiviswanathan et al. showed that the *smooth sensitivity* is bounded by a function of local sensitivity, and any algorithm that is differentially private on $\overline{G}$ can then be transformed into one that is suitable for all graphs [117]. A more efficient method is based on *Lipschitz extension*. A function f' is a *Lipschitz extension* of f from G_θ to G if f' agrees with f on G_θ, and the global sensitivity of f' on G is equal to the global sensitivity of f on

G_υ. Blocki et al. [27] proceeded with a similar intuition. They showed that *Lipschitz extensions* exist for all real-valued functions, and give a specific projection from any graph to a particular degree-bounded graph, along with smooth upper bound on its local sensitivity. Kasiviswanathan et al. [117] proposed efficient constructions for such extensions for *degree distribution* as well as *subgraph counts* such as *triangles*, *k-cycles* and *k-stars*. These techniques can generate statistics with better accuracy for graphs that satisfy an expected condition, but may yield poor results on other graphs. This makes the setting of the degree d a difficult task.

4.4.3   Discussion

Existing methods work reasonably well with edge differential privacy or even node differential privacy guarantee for basic graph statistics. However, releasing specific statistics such as cuts, pairwise distances between nodes, or on hyper-graphs, still remain open issues.

4.5   Summary on Interactive Setting

In interactive settings, the privacy mechanism receives a user's query and replies with a noisy answer to preserve privacy. Traditional Laplace mechanisms can only answer sublinear of n queries, which is insufficient in many scenarios. Researchers have to provide different mechanisms to fix this essential weakness.

The proposed interactive publishing mechanisms improve performance in terms of query type, the maximum number of queries, accuracy and computational efficiency. Upon analysis, we conclude that these measurements in interactive publishings are associated with one another. For example, given a fixed privacy budget, a higher accuracy usually results in a smaller number of queries. On the other hand, with a fixed accuracy, a larger number of queries normally leads to computationally inefficient mechanisms. Therefore, the goal of data publishing mechanism design is to achieve a better result that can balance the above mentioned measurements. In practice, the choice of mechanism depends on the requirement of the application.

Chapter 5
Differentially Private Data Publishing: Non-interactive Setting

Non-interactive settings mean all queries are given to the curator at one time. The key challenge for non-interactive publishing is the sensitivity measurement. Correlation between queries will dramatically increase the sensitivity. Two possible methods are proposed to fix this problem: one is decomposing the correlation between batch queries, which is presented in Sect. 5.1, another is publishing a synthetic dataset with the constraint of differential privacy to answer those proposed queries. Related methods are presented in the synthetic dataset publishing Sections.

5.1 Batch Queries Publishing

Batch queries publishing refers to the most common non-interactive scenario, in which a fixed set of m queries $F(D) = \{f_1(D), \ldots, f_m(D)\}$ are answered in a batch [138]. In this scenario, queries may be correlated to one another, so deleting one record may affect multiple query answers. According to the definition of sensitivity, correlation among m batch queries leads to a higher sensitivity than independent queries.

Table 5.1 shows a frequency dataset D with four variables, and Table 5.2 contains all possible range queries $F = \{f_1, \ldots, f_{10}\}$ that a mechanism will answer. Deleting any record in D will change at most six query results (column containing x_2 in Table 5.1) in F. According to the sensitivity definition in Chap. 2, the sensitivity of F is 6, which is much higher than the sensitivity of a single query. Therefore, most researches focus on how to decrease the sensitivity of F.

© Springer International Publishing AG 2017
T. Zhu et al., *Differential Privacy and Applications*,
Advances in Information Security 69, DOI 10.1007/978-3-319-62004-6_5

Table 5.1 Frequency table D

Grade	Count	Variable
90–100	12	x_1
80–89	24	x_2
70–79	6	x_3
60–69	7	x_4

Table 5.2 m range queries

Range query							
f_1	x_1	+	x_2	+	x_3	+	x_4
f_2	x_1	+	x_2	+	x_3		
f_3		+	x_2	+	x_3	+	x_4
f_4	x_1	+	x_2				
f_5		+	x_2	+	x_3		
f_6					x_3	+	x_4
f_7	x_1						
f_8			x_2				
f_9					x_3		
f_{10}							x_4

5.1.1 Laplace

In this situation, Laplace noise can be added in two different ways but both will introduce high volume of noise. The first method directly adds noise to each query result. The sensitivity of F is $O(n^2)$ and the variance of the noise per answer is $O(n^4/\epsilon^2)$. The second method adds noise to the frequency table, and then generates the range query results accordingly. In this case, the sensitivity is 1, but the noise in F will accumulate more quickly than in the first method. Hence, both methods introduce a large volume of noise and will lead to inaccurate results, while *how to reduce noise* remains a challenge in batch queries release.

5.1.2 Transformation

One possible means of decreasing noise is to re-calibrate the sensitivity. The transformation mechanism maps the original dataset to a new structure, in which the sensitivity of the query set will be adjusted. Noise will be sampled from Laplace distribution according to the adjusted sensitivity. After adding noise, the noisy structure is inverted to generate a new dataset, which is ready to answer the query set. The key issue in the transformation mechanism is to find a suitable structure which has lower sensitivity.

Xiao et al. [237] proposed a wavelet transformation, called *Privelet*, on the dataset to decrease the sensitivity. *Privelet* applies a wavelet transformation on the

Table 5.3 m range queries

Range query							
f_1	x_1	$+$	x_2	$+$	x_3	$+$	x_4
f_2	x_1	$+$	x_2	$-$	x_3	$-$	x_4
f_3	x_1	$-$	x_2				
f_4					x_3	$-$	x_4

frequency dataset D to generate a wavelet coefficient matrix $\mathscr{P}$, in which each entry $p_i \in \mathscr{P}$ is considered as a linear combination of entries in D. *Privelet* adds weighted Laplace noise to each coefficient to create $\mathscr{P}^*$, which is then applied to answer basic queries in F. Other range queries can be generated by the linear combination of these basic queries. Table 5.3 shows the basic queries $\mathscr{P}^*$ could answer. If we provide the query of $range(x_2, x_3) = x_2 + x_3$, the true answer can be generated by $range(x_2, x_3) = 0.5f_1 - 0.5f_3 - 0.5f_4$ and the sensitivity will decrease from 6 to 3. In this way, the sensitivity of the wavelet coefficient is estimated as $1 + \log_2 n$ and the variance of noise per answer is $O((\log_2 n)^3/\epsilon^2)$, which is much smaller results than that in the Laplace mechanism.

Li et al. [138–140] proposed the *Matrix* mechanism which can answer sets of linear counting queries. The set of queries is represented by a matrix A called *workload*, in which each row contains the coefficients of a linear query. Given a set of queries, the *Matrix* mechanism defines a workload A accordingly and obtains noisy answers by implementing the Laplace mechanism. The estimates are then used on the A to generate estimates of the submitted queries.

The essential element of the *Matrix* mechanism is to select A to represent the set of queries. Based on the selection of A, *Matrix* can be extended to a variety of approaches. For example, if A is an identity matrix, *Matrix* can be considered as a normal Laplace mechanism for batch queries. If A is selected by using *Haar wavelet*, it can be extended as *Privelet* [237].

In their recent paper, Li et al. [137] extended the *Matrix* framework in a data aware way. They first partitioned the domain according to the data distribution, then applied the matrix method in the partitioned domain to generate a noisy estimation on dataset. The range query could then be answered using the noisy estimation. This release mechanism diminishes the variation in the query answers. The lower bound on the error of the *Matrix* mechanism is analyzed by their subsequent paper [141], in which they proved that the minimum error achievable for a set of queries is determined by the spectral properties of the queries when they are represented in matrix form.

Similarly, Huang et al. [102] transformed the query sets to a set of orthogonal queries to reduce the correlation between queries. The correlation reduction helps to decrease the sensitivity of query set. Yuan et al. [245] presented a low-rank mechanism (LRM), an optimization framework that minimizes the overall error of the results for a batch of linear queries.

5.1.3   Partition of Dataset

There are other mechanisms of pubishing batch queries, such as partition and iteration. Kellaris et al. [122] decomposed the dataset columns into disjoint groups and added Laplace noise to smooth each group's average count, setting the result as the new count of every column in the group. The final result is generated using the new column's count. Because the maximum number of original counts in a group affects by a user is limited, the sensitivity of each group is decreased and the Laplace noise required for ϵ-differential privacy is also diminishes. The advantage of this mechanism is that it can limit the sensitivity in the numerical dataset.

5.1.4   Iteration

By recursively approximating the true answer, the noise in the output can be effectively diminished. Xiao et al. [235] aimed to decrease the error of the released output. They argued that the Laplace mechanism adds noise with a fixed scale to every query answer regardless value of the answer. Thus, queries with small answers have much higher expected error, which defined as *relative error*. In practice, larger answers can tolerate more noise and for some applications, *relative errors* are more important than absolute errors.

To decrease the relative error, Xiao et al. proposed a mechanism named *iReduct*, which initially obtains rough error estimations of query answers and subsequently utilizes this information to iteratively refine these error evaluations. The algorithm consists of three stages: (1) divides ϵ into two parts and utilizes the first part to perform Laplace mechanism to answer those queries; (2) estimates new noise according to the noise answers; (3) identifies query answers with small but relatively large reduced noise iteratively until errors can be diminished no further. In general, *iReduct* takes advantage of parallel composition, which decreases the errors of the query answers.

5.1.5   Discussion

The key problem in batch query publishing is how to decrease the sensitivity between correlated queries. Currently, transformation is the most popular way to tackle the problem. Current works mainly focused on range queries, and developed appropriate structures to answer linear combination of those queries. More types of structures need to be developed to answer various types of queries. Iteration and partition of dataset may not be effective on the correlation decomposition, but they have the potential to answer more types of queries.

5.2 Contingency Table Publishing

Contingency table is a popular data structure for data analysis in the medical science, and social sciences fields [80]. It displays the frequencies of the combined attributes in a dataset. Suppose D is a frequency dataset of 2^d possible combinations of these attributes. The curator does not normally release the entire contingency table because when d is large, the contingency table is likely to be sparse and noise may outweigh the true answers of queries. Instead, the curator will release subsets containing parts of attributes that are defined as the *k-way marginal frequency* table ($k \leq d$). One contingency table may contain several overlapping marginal frequency tables. Privately answering marginal queries is a special case of counting queries. For example, it may have the form, "*What fraction of individual records in D satisfies certain properties of d_1 and d_2?*"

Table 5.4 shows a contingency table with $d = 4$. Each record r_i has a combination of these four attributes, and there are $2^4 = 16$ possible combinations. When $k = 2$, as shown in the gray area in Table 5.4, the differential privacy mechanism only needs to release $2^2 = 4$ combination of attributes. The chosen of two attributes might be $\{d_1, d_2\}$, $\{d_2, d_3\}$, $\{d_3, d_4\}$ and $\{d_4, d_1\}$. The contingency table release not only requires the privacy of *k-way marginal frequency* tables, but also the consistency of these tables. For example, the counts of the same attributes in different marginal frequency tables should be equal and should not violate common senses.

5.2.1 Laplace

Two methods exist to release *k-way marginal frequency* tables, both of which involve the direct implementation of the Laplace mechanism. The most intuitive method adds noise into the frequency of the whole contingency table [118]. Users can create any *k-way marginal frequency* table from the noisy contingency table and maintain the consistency of all tables. However, this method leads to noise magnitude of $O(2^d)$. When d is large, the noise will increase dramatically and renders the perturbation results unrealistic.

The other method is to extract the marginal frequency tables from the original dataset and add the noise to the marginal frequencies. This method achieves excellent accuracy when n is large compared to the number of marginal tables,

Table 5.4 Contingency table with 2-way marginal frequency

Index	d_1	d_2	d_3	d_4	Count
$index_1$	0	0	0	0	4
$index_2$	0	1	0	1	3
$index_3$	1	0	1	0	3
...	...	...	...	...	...
$index_{16}$	0	1	1	1	1

but it can lead to the violation of consistency. For example, after adding noise, the maximum score may be lower than the average score. Based on the weaknesses of the traditional Laplace mechanism, researchers have attempted to find new solutions to the accuracy and consistency problems. The most popular solutions are using the iteration and transformation mechanisms.

5.2.2 Iteration

Ding et al. [58] proposed a differentially private data cubes to generate marginal tables. The proposed method first organize all possible marginal frequency tables in a lattice. With a d-dimensional binary dataset, there are 2^d 2-ways marginal frequency tables in the lattice. The method then iteratively selects which tables should be released. In each iteration, it traverses all 2^d tables in the lattice and greedily selects one. Obviously, this method requires time polynomial in 2^d and is impractical. Qardaji et al. [186] improved the above method in a practical way. Instead of generating 2^d tables, they choose sets of attributes and use them as a synopsis from which one can reconstruct any desired k-way marginal frequency tables.

To improve accuracy, Chandrasekaran et al. [35] proposed an iterative dataset construction mechanism, which maintains a sequence of queries, $f_1(D), f_2(D), \ldots,$ that give increasingly improved approximations to the $F(D)$. The algorithm is capable of answering marginal queries with a worst-case accuracy guarantee for dataset containing $poly(d, k)$ records in time $exp(o(d))$.

5.2.3 Transformation

Similar to histogram publishing, contingency table publishing also meets with the problem of inconsistency. Barak et al. [14] proposed an algorithm to retain consistency by transforming the contingency table into a *Fourier* domain. This transformation serves as a non-redundant information encoding method for marginal tables. Since any set of *Fourier* coefficients corresponds to a contingency table, adding noise in this domain does not violate the consistency. Linear programming is applied to obtain a non-negative, integrated marginal frequency table with the given *Fourier* coefficients. Without compromising the differential privacy, a set of consistent marginal frequency tables was produced.

5.3 Anonymized Dataset Publishing

The anonymized dataset publishing retains the format of the original records. This is an attractive property as in some scenarios, people need to know the details of attributes to determine further analysis methods. Data release can not achieve this goal as users are only allowed to access the dataset by submitting queries. The anonymized dataset publishing can guarantee that the published records have the same attributes with the original records. The problem of anonymized dataset publishing can be stated as follows: suppose the input dataset is D with d attribute d_i, the curator would like to publish a $\widehat{D}$ with the same attributes.

The anonymized dataset publishing has a long research history and has been investigated heavily in the privacy-preserving community. Differentially private anonymized dataset publishing, however, is a tough problem as publishing a specific record is considered to be the violation of differential privacy. Researchers meet with the difficulty that on the one hand, they have to release specific records. On the other hand, they should meet the requirement of differential privacy.

To address the difficulty, Mohammed et al. [161] observed that if the anonymization process follows the requirement of differential privacy at each step, the result will satisfy with differential privacy. Based on the observation, they proposed an anonymized algorithm *DiffGen* to preserve privacy for data mining purpose. The anonymized procedure consists of two major steps: *partition* and *perturbation*. Every attribute of the original dataset is generalized to its topmost state. The partition step then splits these attributes into more specific groups according to attribute taxonomy trees. It applies an exponential mechanism to select a candidate for specialization. After that, the perturbation step adds noise to the true count of each records group.

Table 5.5 shows an example on a transaction dataset. To release this dataset, *DiffGen* generalize each attribute to its topmost according to pre-defined taxonomy trees, which are shown in Fig. 5.1. Specifically, for the attribute *Job*, Fig. 5.2 shows that all types of jobs are generalized to *Any__Job* and then partitioned according to the job taxonomy tree. Similarly, all ages are generalized to [18–60) and then are split into [18–40) and [40–65). In Fig. 5.2, the root of the partition tree is the count of full generalized records (Age = [18–65), Job = Any__Job). Let the

Table 5.5 Transaction dataset

Job	Age	Class
Engineer	34	Y
Lawyer	50	N
Engineer	38	N
Lawyer	33	Y
Dancer	20	Y
Writer	37	N
Writer	32	Y
Dancer	25	N

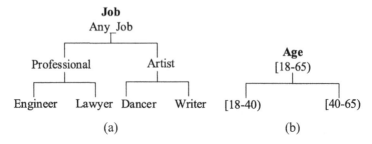

Fig. 5.1 Taxonomy tree examples. (**a**) Taxonomy tree of job. (**b**) Taxonomy tree of age

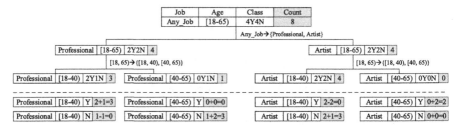

Fig. 5.2 Anonymized dataset example

Table 5.6 Set-value dataset

TID	Items
t_1	$\{I_1, I_2, I_3, I_4\}$
t_2	$\{I_2, I_4\}$
t_3	$\{I_2\}$
t_4	$\{I_1, I_2\}$
t_5	$\{I_2\}$
t_6	$\{I_1\}$
t_7	$\{I_1, I_2, I_3, I_4\}$
t_8	$\{I_2, I_3, I_4\}$

first specialization be *Any_Job→Professional, Artist*, the *DiffGen* creates two new partitions under the root and splits data records between them. The partition is repeated until records are specified to a predefined level. Laplace noise is then added to the count of each records group.

The *DiffGen* is the initial work that bridges the gap between anonymization and the differential privacy. It is an efficient algorithm with the time complexity of $O(n \log n)$. In addition, because the publishing is for data mining purpose, the accuracy is measured by the performance of the decision tree.

Similarly, Chen et al. [43] follows the same strategy to deal with another type of transaction dataset, set-value data. It refers to the data in which each record is associated with a set of items drawn from a universe of items. Table 5.6 illustrates an example of set-value data.

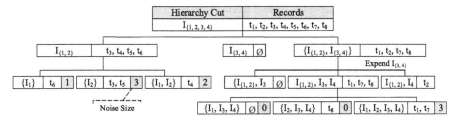

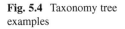

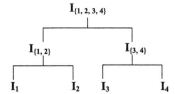

Fig. 5.3 Anonymized set-value dataset example

Fig. 5.4 Taxonomy tree
examples

Chen et al. proposed *DiffPart* to publish the set-value data. The difference between *DiffPart* and *DiffGen* is that the *DiffPart* partitions the dataset based on a context-free taxonomy tree while *DiffGen* utilizes the taxonomy tree based on the underground dataset. Figure 5.3 shows an example of *DiffPart*, and Fig. 5.4 is the taxonomy tree of the itemset.

The *DiffPart* has the computational time complexity of $O(n \cdot |I|)$. It publishes dataset for frequent mining purpose, so the accuracy is measured based on the accuracy of frequent mining.

5.4   Synthetic Dataset Publishing

The synthetic dataset follows the distribution of the original dataset but not necessary share the same format. It was considered as a difficult problem for a long time due to the large noise introduced. It leads to an inaccurate output. In this perspective, publishing a synthetic dataset by using Laplace mechanism is hard.

5.4.1   Synthetic Dataset Publishing Based on Learning Theory

5.4.1.1   Learning Theory in Differential Privacy

Learning theory is one of the main tools used in differential privacy. The most prevalent concept is *Probably Approximately Correct* (PAC) learning. It helps to measure the lower bound on *sample complexity* and *accuracy*, both of which are widely used in differential privacy to evaluate the performance of private algorithms.

Suppose a set of samples D drawn from a universe $\mathscr{X}$. The samples are labeled according to a concept $c : \mathscr{X} \rightarrow \{0, 1\}$ from a concept class $\mathscr{C}$. The goal of the learning process is to find a hypothesis $h : \mathscr{X} \rightarrow \{0, 1\}$ which agrees with c on almost the entire universe. To quantify the accuracy guarantee α of the learning algorithm, a lower bound of sample complexity is provided to evaluate how many samples in $\mathbf{r}$ are needed to guarantee an accuracy of α.

A concept $c : \mathscr{X} \rightarrow \{0, 1\}$ is a predicate that labels samples by either 0 or 1. A group of concepts is defined as a concept class $\mathscr{C}$. The learning algorithm is successful when it outputs a hypothesis h that approximates the target concept c over samples from D. More formally,

Definition 5.1 (PAC Learning) Algorithm $\mathscr{A}$ is a PAC learner for a concept class $\mathscr{C}$ over $\mathscr{X}$ using hypothesis class H if for all concepts $c \in \mathscr{C}$, all distributions $\mathscr{D}$ on $\mathscr{X}$, given an input of n samples and given that each r_i is drawn i.i.d. from D, algorithm $\mathscr{A}$ outputs a hypothesis $h \in H$ satisfying

$$Pr[error_D(c, h) \leq \alpha] \geq 1 - \beta. \tag{5.1}$$

The probability is taken over the random choice of the samples in D.

The concept quantifies the guarantee of an algorithm in terms of a lower bound on the dataset size. The bound on the dataset size is the *sample complexity* of an algorithm and α is the *accuracy*. These two measurements on algorithms are corresponding to each other and are considered to be basic utility measurements in differentially private data release, publishing and analysis.

Kasiviswanathan et al. [115] defined a private PAC learning as a combination of PAC learning and differential privacy. Wang et al. [225] confirms the result that a problem is privately learnable if there is a private algorithm that asymptotically minimizes the empirical risk. These theories build a bridge between learning theory and differential privacy, so that these traditional learning theories can be applied in data release, publishing and analysis.

5.4.1.2 Synthetic Publishing

With the theoretical development in differential privacy, the inaccuracy problem can be partly solved by introducing learning theory to the synthetic dataset release. The findings of the principal paper by Kasiviswanathan et al. [115] ensured that almost anything learnable can be learned privately. In addition, Blum et al. [29] subsequently claimed that the main purpose of analyzing a dataset is to obtain information about a certain concept. If the query on a published dataset is limited to a particular concept, the learning process can ensure the accuracy of the query output.

Kasiviswanathan et al. and Blum et al.'s results linked the synthetic dataset with the learning process. The query set F on synthetic dataset are associated with *Boolean* predictions of a concept set $\mathscr{C}$. A synthetic $\widehat{D}$ is considered as a hypothesis

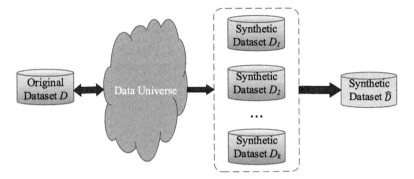

Fig. 5.5 Synthetic dataset release

that satisfies $\mathscr{C}$. For every concept c derived from $\mathscr{C}$, we consider it as a query and the distance between the query result in $\widehat{D}$ and D is limited by an accuracy parameter α. The release target is to search for a good hypothesis $\widehat{D}$ fitting for $\mathscr{C}$. The publishing objects are (1) to find a tight bound on the accuracy α, and (2) to release the result efficiently.

Accuracy

Kasiviswanathan et al. [115] proposed an exponential based mechanism to search a synthetic dataset from the data universe that can accurately answer $\mathscr{C}$. The search process is based on the assumption that the original dataset and the synthetic dataset are sampled from a same universe $\mathscr{X}$. Figure 5.5 shows that after creating multiple candidate datasets from the data universe, the mechanism will search the most suitable $\widehat{D}$ based on the exponential mechanism. The authors claimed that for any class of $\mathscr{C}$ and any dataset $D \geq \{0, 1\}^d$, if the size of the dataset satisfies

$$n \geq O\left(\frac{dVC(\mathscr{C})\log(\frac{1}{\alpha})}{\epsilon\alpha^3} + \frac{\log\frac{1}{\delta}}{\epsilon\alpha}\right),$$

α accuracy can be successfully achieved. The mechanism is only suitable for discrete domain.

Blum et al. [29] applied a similar exponential based mechanism, *net* mechanism, to generate a synthetic dataset over a discrete domain. They searched for several net datasets $N(\mathscr{C})$ that could answer queries from $\mathscr{C}$. The net mechanism provides a lower bound of the accuracy:

$$\alpha \geq O\left(\frac{n^{2/3}\log^{1/3}m\log^{1/3}|\mathscr{X}|}{\epsilon^{1/3}}\right).$$

Subsequent works made progress in proving accuracy. Dwork et al. [74] utilized *boosting* to improve the accuracy of the release dataset and improved the accuracy lower bound to $O(\frac{\sqrt{n \log |\mathcal{X}|} \log^{2/3} m}{\epsilon})$. Hardt et al. [91] combined the exponential mechanism with a multiplicative weight updating approach to achieve a nearly optimal accuracy guarantee.

Efficiency

Although these mechanisms provide a tight bound on accuracy, none of them can be performed in polynomial time. The time cost is super-polynomial in the size of universe $\mathcal{X}$ and the concept class $\mathcal{C}$. It is $|\mathcal{X}|^{poly(n) \log |\mathcal{C}|}$. Blum et al. [29] claimed that if a polynomial time is required, the definition of privacy has to be relaxed. This claim was confirmed by Dwork et al. [71], who proved that only after relaxing the notion of ϵ to (ϵ, τ) could the run-time be linear in the size of the data universe and the size of the query set.

Moreover, Ullman et al. [221] showed that no algorithm can answer more than $O(n^2)$ arbitrary predicate queries in polynomial time. The Laplace mechanism is almost optimal among all computationally efficient algorithms for privately answering queries. This result suggests that for privately query release, it is difficult to design mechanisms to answer arbitrary queries efficiently. Classes of queries that have a particular structure is what we can exploit.

Gaboardi et al. [85] presented a practical way to deal with the inefficient problem for high dimensional datasets. The algorithm packages the computationally hard step into a concisely defined integer program, which can be solved non-privately using a standard solver. The optimization step does not require a solution that is either private or exact: it can be quickly solved by existing, off-the-shelf optimization packages quickly in practice. Even though the authors do not solve the inefficient problem in synthetic data release, their work provides a practical way to compute high dimensional data in practice.

Discussion

Table 5.7 summarizes key works on the synthetic dataset in terms of accuracy, efficiency and privacy level. For simplicity, sensitivity is predefined as 1, the dependence on β is omitted and the run-time only considers the size of the universe.

In summary, computational learning theory extends the research work on synthetic data release, proving it is possible to maintain an acceptable utility while preserving differential privacy. Nevertheless, the issues of *how to reduce the computational complexity*, and *how to provide various types of queries on these datasets* remain as challenges.

Table 5.7 Synthetic data release comparison

Mechanism	Accuracy	Efficiency	Privacy
Net mechanism [29]	$O(\frac{n^{2/3}\log^{1/3}m\log^{1/3}\mathscr{X}}{\epsilon^{1/3}})$	$\exp(\mathscr{X})$	(ϵ)
DNR [71]	$O(\frac{n^{1/2}\log^{1/2}\mid\mathscr{X}\mid m}{\epsilon})$	$\exp(\mathscr{X})$	(ϵ,δ)
DRV [74]	$O(\frac{n^{1/2}\log^{1/2}\mid\mathscr{X}\mid\log^{3/2}m}{\epsilon})$	$\exp(\mathscr{X})$	(ϵ,δ)
MWP [91]	$(\frac{\log\mid\mathscr{X}\mid\log m}{\epsilon n})^{1/3}$	$\exp(\mathscr{X})$	ϵ

5.4.2 High Dimensional Synthetic Dataset Publishing

Neither anonymized dataset publishing nor learning theory based publishing can effectively handle the high-dimensional dataset. For the anonymized method, when the input dataset contains many attributes, existing anonymized method will inject a prohibitive amount of noise, which leads to an inferior utility. For the learning theory based method, the computational complexity is exponential to the dimension of the dataset, making the publication infeasible to high-dimensional dataset. One promising way to address the high dimensional problem is to decompose high dimensional data into a set of lower dimensional marginal datasets, along with an inference method that infers the joint data distribution from these marginal datasets.

Zhang et al. [248] followed the above rationale and applied Bayesian network to deal with the high dimensional problem. They assumed that there are correlations between attributes. If these correlations can be modeled, the model can be used to generate a set of marginals to simulate the distribution of the original dataset. Specifically, given a dataset D, they first construct a Bayesian network to model the correlations among the attributes in D. This model will approximate the distribution of data in D using a set of low-dimensional marginals of D. After that, they inject noise into each marginal to ensure differential privacy. These noisy marginals and the Bayesian network are used to build an approximation of the data distribution of D. Lastly, they sample records from the approximate distribution to construct a synthetic dataset. The disadvantage of the solution is that it consumes too much privacy budget in the Bayesian network constructing process, making the approximation of the distribution inaccurate.

Chen et al. [44] addressed the disadvantage by proposing a clustering based method. They first learn the pairwise correlation of all attributes and generate a dependency graph. Secondly, they apply the junction tree algorithm to the dependency graph to identify a collection of attribute clusters to derive all the noisy marginals. At last, they make use of the noisy marginal tables and the inference model to generate a synthetic dataset. Comparing with [248], they have limited access to the dataset in the two steps, saving the privacy budget to obtain a better result in the last step.

5.5 Summary on Non-interactive Setting

The interactive setting has attracted attention due to advances in statistical databases. In interactive settings, the privacy mechanism receives a user's query and replies with a noisy answer to preserve privacy. Traditional Laplace mechanisms can only answer $O(n)$ queries, which is insufficient in many scenarios. Researchers have to provide different mechanisms to fix this essential weakness.

The proposed interactive publishing mechanisms improve performance in terms of query type, the maximum number of queries, accuracy, and computational efficiency. Upon analysis, we conclude that these measurements in interactive releases are associated with one another. For example, given a fixed privacy budget, a higher accuracy usually results in a smaller number of queries. On the other hand, with a fixed accuracy, a larger number of queries normally leads to computationally inefficient mechanisms. Therefore, the goal of data publishing mechanism design is to achieve a better result that can balance the above mentioned measurements. The choice of mechanism depends on the requirement of the application.

High sensitivity presents a big challenge in non-interactive settings. Batch query publishing methods can only publish limited types of queries. Publishing a synthetic dataset seems appealing because, in some scenarios, people require details of the attributes to determine further analysis methods. The research on synthetic data publishing, however, is still in its early stages and there are many open problems in this area. The essential problem is efficiency. Given most publishing mechanisms need to sample datasets from the entire data universe, it is hard to search for a suitable dataset in polynomial time.

Another problem is that synthetic dataset publishing can only publish datasets for particular purposes. For example, an anonymization dataset focuses on a decision tree algorithm, the published dataset obtains an acceptable result for decision tree tasks, yet the proposed method does not guarantee the performance for other types of tasks. Learning-based methods have the same disadvantage, or worse, as they only guarantee learning performance for a particular class. Publishing a dataset for multiple purposes needs further investigation.

The third problem is dealing with high-dimensional datasets. Even though [248] and [44] have undertaken some initial work, they both consume too much of the privacy budget when building the distribution model, making the results less accurate than that of lower-dimensional datasets.

Chapter 6
Differentially Private Data Analysis

Privacy is an increasingly important issue in a wide range of applications, and existing non-private algorithms need to be improved to satisfy the requirement of privacy preserving. These non-private algorithms can be data mining, machine learning or statistical analysis algorithms. The straightforward way to preserve the privacy is to incorporate the differentially private mechanisms, such as Laplace or exponential mechanism, into existing algorithms. We consider this incorporation as *differentially private data analysis*, which is an important research direction that attracts significant research attention.

Differentially private data analysis aims to publish an approximately accurate analysis model rather than query answers or synthetic datasets. The essential idea is to extend the current non-private algorithm to a differential privacy algorithm. This extension can be realized by several frameworks, which can be roughly categorized as a *Laplace/exponential framework* and a *Private Learning Framework*. Table 6.1 shows the differentially private data analysis problem characteristics, in which researchers concerned with accuracy and computational efficiency of these two frameworks. As different papers use diverse terms to describe the output, terms "model" and "algorithm" are interchangeable in this section.

6.1 Laplace/Exponential Framework

The most common extension method is to incorporate Laplace or exponential mechanisms into non-private analysis algorithms, and hence it is referred to as Laplace/exponential framework.

© Springer International Publishing AG 2017   49
T. Zhu et al., *Differential Privacy and Applications*,
Advances in Information Security 69, DOI 10.1007/978-3-319-62004-6_6

Table 6.1 Differentially private data analysis problem characteristics

Differentially private data analysis	
The nature of input data	Transaction
The nature of output data	Analysis models/algorithms
Analysis framework	*Laplace/exponential* framework, private learning
Analysis mechanism	*Laplace/exponential*, learning process
Challenges	Accuracy, computational efficiency

Fig. 6.1 SuLQ interface

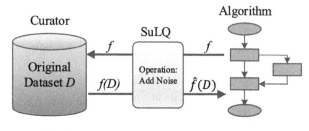

Fig. 6.2 PINQ interface

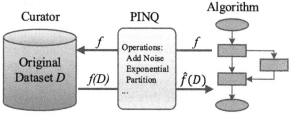

6.1.1   SuLQ and PINQ Interface

As not every analyzer is an expert in privacy, researchers have developed a number
of standard interfaces that automatically bring differential privacy into non-private
algorithms. Two essential interfaces satisfied the Laplace/exponential framework:
Sub-Linear Queries (SuLQ) interface [28] and *Privacy Integrated Queries* (PINQ)
interface [155]. Both assume that non-private algorithm can be decomposed into
operations, which can be considered to be queries set F. Each operation $f \in F$ is
submitted to the interfaces, and the interfaces replied with randomized results in
the differential privacy constraint. After combining these randomized query results,
the algorithm provides a differentially private result. Figures 6.1 and 6.2 shows two
interfaces.

6.1.1.1   SuLQ

Sub-Linear Queries (SuLQ) output randomized continuous and Boolean values
for each input. It enforces the differential privacy by adding calibrated noise into
each operation. The analyzer utilizes these noisy operations to generate more
sophisticated algorithms.

One of the limitations of SuLQ lies in its lack of the exponential mechanism. When dealing with selection operation of an algorithm, SuLQ is unable to provide sufficient privacy guarantees [83]. Therefore, SuLQ needs further improvement to satisfy diverse algorithms.

6.1.1.2 PINQ

Privacy Integrated Queries platform (PINQ) [155] is an interface that provides more operations than SuLQ. It uses Laplace noise on numeric queries and exponential mechanism on selection operations. It not only defines the numeric queries such as count (NoisyCount), sum (NoisySum) and average (NoisyAvg) with independent noise, but also provides an operation Partition that allows queries to execute on disjoint datasets. The Partition operation takes advantage of parallel composition by partitioning the dataset into multiple disjoint sets. Its privacy level is determined by the maximal ϵ in these sets. Recently, Proserpio et al. [184] extended the platform into wPINQ (for weighted PINQ), which uses weighted datasets to avoid the worst case sensitivities.

In summary, based on these private operations, SuLQ and PINQ can create private algorithms such as *ID3* classifier, *singular value decomposition, k-means*, and statistical query learning models [28]. However, as these interfaces do not consider the objectives and properties of various algorithms, the performance of the analyzed results may be suboptimal.

6.1.2 Specific Algorithms in the Laplace/Exponential Framework

We list some examples of algorithms to illustrate the implementation of the Laplace/Exponential framework. Some of them apply an interface such as SuLQ, while some introduce the Laplace and exponential mechanisms into the algorithm directly. These algorithms are usually associated with specific machine learning or data mining tasks, which are categorized into *Supervised Learning, Unsupervised Learning* and *Frequent Itemset Mining*.

6.1.2.1 Supervised Learning

Supervised learning refers to the prediction methods that extract models describing data classes via a set of labeled training records [108]. As one of the most popular supervised learning algorithms, *decision tree* learning, has been extensively studied in Laplace/exponential frameworks.

Decision trees are iterative processes that recursively partition the training sample to build a tree with each label representing a leaf. Assuming there is an input dataset D with d categorical attributes $\{a_1, \ldots a_d\}$, a decision tree is constructed from the root that holds all the training records then the algorithm chooses the attribute a_i that maximizes the information gain to partition the records into child nodes. The procedure is performed recursively on each subset of the training records until a stop criteria is meet.

The first differentially private decision tree algorithm was developed on the SuLQ platform [28]. Noise is added to the information gain, and an attribute a_i with noisy information gain that is less than a specified threshold is chosen to partition a node. However, as information gain is evaluated separately for each attribute in each iteration, the privacy budget is consumed several times in each iteration, which results in a large volume of noise. In addition, SuLQ fails to deal with continuous attributes. If those attributes are simply discretized into intervals, the basic concept of differential privacy is violated, because the split values in continuous attributes would reveal information about the records.

To overcome the SuLQ platform's disadvantages, Friedman et al. [83] improved the algorithm in two ways. First, they implemented an exponential mechanism in the attribute selection step. The score function is defined by the information gain or the gain ratio. The attributes with a top score have a higher probability of being selected. In this way, less of the privacy budget is consumed than by SuLQ. Second, the proposed method can deal with continuous attributes. An exponential mechanism is employed to select every possible splitting value and the continuous attribute's domain is divided into these intervals. Compared to SuLQ, they obtain better performance.

Jagannatham et al. [104] provided an algorithm for building random private decision trees, which randomly select attributes to create nodes. The algorithm first creates a tree in which all the leaves are on the same level and then builds a leaf count vector. Once the independent Laplace noise is added to the count vector, a differentially private random decision tree can be generated from the noisy leaf vector. The algorithm iteratively produces multiple random decision trees and uses the ensemble method to combine these trees. As the attribute is randomly selected, this step saves the privacy budget; however, as each tree's magnitude is scaled up with the number of trees in the final ensemble step, the utility of the ensemble remains a problem.

Rana et al. [188] proposed a practical approach to ensemble decision trees in a random forest. They do not strictly follow the notion of differential privacy, which keeps the neighboring data distribution approximately invariant. Instead, they only keep the statistic features invariant. A privacy attack model is defined to prove the reliability of the proposed decision tree. As less budget has been consumed in the ensemble process, this relaxation of differential privacy can lead to higher utility compared to other algorithms.

Discussion Differentially private decision tree algorithms were the earliest algorithms to be investigated in the differential privacy community. Table 6.2 compares

Table 6.2 Comparison of supervised learning methods

Difficulty	Key methods	Typical papers	Advantages	Disadvantages
Privacy budget has to be consumed multiple times	Add noise to information gain	[28]	Easy to implement	Noise will be high due to privacy budget arrangement in attribute selection
	Use exponential mechanism to select attribute	[83]	Save part of privacy budget in the attribute selection	Still has high noise
	Select attribute randomly	[104, 188]	Does not consume privacy budget during attribute selection	Privacy budget will be largely consumed in the ensemble process

most prevalent differentially private methods in decision trees. The advantage of this series of methods is that they are concise and easy to implement. However, because tree-based algorithms need to select split attributes multiple times, the privacy budget is quickly consumed, which incurs a huge utility loss. This drawback stems from decision tree building and is not easy to deal with. Nowadays, one of the most popular ways to design differentially private supervised learning algorithms is to apply a private learning framework.

Algorithm 1 Basic k-means Algorithm

Require: points $r_1, \ldots, r_n$.
Ensure: k centers $\{v_1, \ldots, v_k\}$ with arranged points
   repeat
      1. Create k clusters by assigning each point to its closest center;
      2. Re-calculate the center of each cluster;
   until centers do not change

6.1.2.2 Unsupervised Learning

As a typical unsupervised learning method, *clustering* algorithms group unlabeled records into clusters to ensure that all records in the same cluster are similar to each other. Assume the input is a set of points, $r_1, \ldots, r_n$, the clustering output is k centers $\{v_1, \ldots, v_k\}$ and assigned points.

A basic k-means algorithm is formulated by Algorithm 1. The aim of differential privacy clustering is to add uncertainty into the center v and the number of records in each center. To achieve this goal, noise is added in Step 1 in Algorithm 1. In fact, adding noise to cluster centers is impractical because the sensitivity of the cluster center will be quite large as deleting one point will totally change the center.

Therefore, the challenge for clustering is to evaluate and minimize the sensitivity of the cluster centers.

Nissim et al. [172] used local sensitivity with k-means clustering to circumvent the large sensitivity problem, by relying on the following intuition. In a well-clustered scenario, a point with noise should have approximately the same center as its previous center. In addition, moving a few "well-clustered" records would not ultimately change the centers. As such, they define a local sensitivity to measure the record-based sensitivity of the cluster center, which is much lower than the traditional global sensitivity. Since the value of local sensitivity is difficult to measure, they provide a sample aggregate method to approximate local sensitivity.

Feldman et al. [79] proposed a novel private clustering by defining the private *coresets*. A coreset P is a small weighted set of records that captures some geometric properties of these records. They use the private coreset to preserve differential privacy for k-median queries. When calculating what is the sum of distances from P to some other records in set R, the input set R can be transferred under the boundary of differential privacy to a private coreset P, in which k-median queries can easily be handled. Because the coreset P is differentially private, the clustering algorithm performed on P also satisfies differential privacy.

Based on local sensitivity, Wang et al. [227] implemented subspace clustering algorithms. They introduce Laplace mechanism into an agnostic k-means clustering, and an exponential mechanism into Gibbs sampling subspace clustering algorithm. Their subsequent paper, Wang et al. [228] adopted a Johnson-Lindenstrauss transform to guarantee differential privacy in a subspace clustering algorithm. The Johnson-Lindenstrauss transform can reduces the dimensions of the dataset, making the clustering problem practical, as well as preserving the distance between each record.

Discussion In general, even some differentially private platforms such as SuLQ [28], PINQ [155], Privgene [251], Gupt [162] can automatically implement clustering algorithms. They all assume that the sensitivity of the cluster centers has been predefined. Even though local sensitivity can partially solve the problem, further reducing sensitivity still remains a challenge. Table 6.3 illustrate current differentially private unsupervised methods.

6.1.2.3   Frequent Itemset Mining

Frequent itemset mining aims to discover itemsets that frequently appear in a dataset D. Suppose I is the set of items in D, and an *itemset* refers to a subset of I. Let each record $r_i \in D$ denote as a transaction that contains a set of items from I. A frequent itemset refers to a set of items whose number of occurrences in transactions is above a threshold, and the proportion of supporting transactions in the dataset is defined as the *frequency*. Given an itemset I_j, if transaction r_i contains I_j, we say r_i supports I_j, and the proportion of supporting transactions in the dataset is defined as the *support*

Table 6.3 Comparison of unsupervised learning methods

Difficulty	Key methods	Typical papers	Advantages	Disadvantages
High sensitivity of clustering centers	Use local sensitivity	[172, 227]	Decreases the level of the sensitivity	Local sensitivity may not be easy to estimate
	Johnson-Lindenstrauss transform	[228]	Guarantees differential privacy while retaining distance between points	Johnson-Lindenstrauss transform is only valid for norm 2 distance measurement

of I_j. An itemset with a frequency larger than the predefined support threshold is called a frequent itemset or a frequent pattern.

Let U represent all frequent itemsets, where the *topk* most frequent itemsets in U should be released under differential privacy guarantee. Laplace noise is normally added to the frequency; however, the main challenge is that the total number of itemsets is exponential to the number of items: If I contains n items, the number of all possible itemsets is $|U| = \sum_{i=1}^{k} \binom{n}{i}$. Decreasing the number of candidate itemsets is a major research issue in differentially private frequent itemset mining.

Bhaskar et al. [24] solved the problem by providing a truncated frequency to reduce the number of candidate itemsets. They proposed an algorithm that uses the exponential mechanism to choose the top-k itemsets. The score function of each candidate is the frequency defined as $\widehat{p}(U) = max(p(U), p_k - \gamma)$, where p_k is the frequency of the k-th most frequent itemsets and $\gamma \in [0, 1]$ is an accuracy parameter. Every itemset with a frequency greater than $p_k - \gamma$ is computed as its normal frequency $p(U)$ while the frequency of the rest is truncated to $p_k - \gamma$. All the itemsets with frequencies of $p_k - \gamma$ are grouped into one set, and the algorithm uniformly selects itemsets from this set. In this way, the computational cost is significantly reduced.

The advantage of the *truncate frequency* is that it could significantly decrease the size of candidate itemsets. However, it is only applicable when k is small, otherwise, the accuracy parameter γ might be larger than q_k. Another weakness is that the *top-k* itemsets should be predefined as length, which affects the flexibility of the frequent itemset mining.

To address these weaknesses, Li et al. [144] proposed an algorithm, *PrivBasis*, which introduces a new notion of *basis sets* to avoid the selection of *top-k* itemsets from a very large candidate set. Given some minimum support threshold, θ, the authors construct a basis set $\mathscr{B} = B_1, B_2, \ldots$, so that any itemset with a frequency higher than θ is a subset of basis B_i. They introduce techniques for privately constructing basis sets, and for privately reconstructing the frequencies of all subsets of B_i with reasonable accuracy.

The *PrivBasis* algorithm releases arbitrary length itemsets and ensures high accuracy, but generating *basis sets B* is not easy. Furthermore, when the length of itemset l and the number of *basis sets* w are large, the cardinality of the candidate set is still too big to handle. Hence, how to efficiently decrease the size of the candidate set is still a challenge.

Zeng et al. [246] utilized a random truncated method to decrease the number of candidate itemsets. They propose an algorithm that randomly truncates transactions in a dataset according to a predefined maximal cardinality. The algorithm then iteratively generates candidate itemsets by the a priori property, and perturbs the support of candidate itemsets.

Lee et al. [134] proposed a FP-tree based frequent itemset mining algorithm. The propose algorithm first identify all frequent itemsets. The algorithm does not know their supports, but only that their supports are above the threshold. Using this information, the algorithm builds a differentially private FP-tree with privatized supports of the frequent itemsets. The proposed algorithm injects noise in the data structure at the intermediate (FP-tree) step. The final output can be further refined through an optional post-processing step. The advantage of the algorithm is that information disclosure affecting differential privacy occurs only for count queries above the threshold; negative answers do not count against the privacy budget.

Shen et al. [200] applied a Markov chain Monte Carlo (MCMC) sampling method to deal with the challenge of large candidate itemsets. They claim that an MCMC random walk method can bypass the problem, and the exponential mechanism can then be applied to select frequent itemsets. They use this method to mine frequent graph patterns.

Xu et al. [244] applied a binary estimation method to identify all possible frequent itemsets and then use an exponential mechanism to privately select frequent itemsets. The number of candidates is eventually a logarithm of the original number using a binary search.

Discussion Frequent itemset mining is a typical data mining task, which suffers from searching large candidate sets. Differential privacy makes the problem worse.

Table 6.4 Comparison of frequent itemset mining methods

Difficulty	Key methods	Typical papers	Advantages	Disadvantages
Large candidate itemsets	Merge candidates into groups	[24, 134, 144, 246]	Easy to implement	Merge strategy has high impact on the output. Some important patterns may be missed
	Binary search in the candidate set	[244]	The frequent itemsets can be quite accurate	The search time can be decreased, but still inefficient
	Sampling from candidate set	[200]	Highly efficient	Some important patterns may be missed

Table 6.4 lists current key methods, in which [24, 134, 246] and [144] tried to merge some candidates into groups using different methods. [200] adopted a sampling method to search the candidate itemsets. [244] applied a binary search method. All of those methods decreased the searching space from exponential to polynomial, which is a big achievement. However, the noise still remains high, which needs to be decreased further.

6.1.3 Summary on Laplace/Exponential Framework

The Laplace/exponential frameworks can introduce Laplace or exponential mechanisms freely to various types of algorithms. However, the utility of the analysis results is a challenge for this framework. When adding noise to algorithm steps, it is unclear how large the utility loss will be and the analysis result is not easy to compare.

One possible way to tackle the difficulty is solving the data analysis problem through the view of optimization and taking advantage of some existing theories, such as the learning theory [115], so that the utility loss can be estimated and compared. Based on this intuition, researchers proposed the private learning framework.

6.2 Private Learning Framework

Machine learning is one of the most commonly-used techniques in various areas. With the increased concern about the privacy preservation, many learning algorithms attempt to protect the privacy of individuals in the training datasets. One line of research investigates the learning problem from the perspective of machine learning theory [115, 225].

A non-private learning problem is shown in Fig. 6.3a. Suppose $D = r_1, \ldots, r_n$ is a set of samples (training dataset) drawn from a universe $\mathscr{X}$. Dots and triangles denote the labels $y \in 0, 1$. A concept c is a function that separates one label from another. Suppose there are groups of functions (hypotheses) H and h_j is one of them. The goal of the learning process L is to find a hypothesis h that agrees with c on almost the entire universe. Figure 6.3b illustrates a typical learning process: by using the input training samples, the learner selects the most suitable $h \in H$ as the output model $\mathbf{w}$.

Based on the description of non-private learning, the purpose of a private learning framework is to design a learner that outputs an approximately accurate model and preserves the differential privacy of training samples. The learner, known as a private learner, preserves the privacy of the training samples and outputs a hypothesis with bounded accuracy. Private learning frameworks are concerned with the following issues:

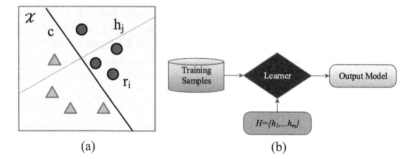

Fig. 6.3 Learning (**a**) problem and (**b**) process description

- *How to choose an optimal model in terms of differential privacy?*
- *How many samples are needed to achieve a bounded accuracy?*

The first question can be dealt with by analyzing the property of Empirical Risk Minimization (ERM) technique, which will be introduced in Sect. 6.2.1. The second question can be tackled by analyzing the sample complexity of a private learner, which will be discussed in Sect. 6.2.3.

6.2.1   Foundation of ERM

The key technique utilized to select a hypothesis is ERM, in which an optimal model is chosen by minimizing the expected loss over the training samples [38].

Suppose $h \in H$ is a hypothesis and $\mathbf{w}$ is the output model, we define a *loss function* $\ell(h(\mathbf{w}, r), y)$ to estimate the expected risk of the hypothesis. The goal of ERM is to identify a $\mathbf{w}$ that minimizes the expected risk of the whole universe. Equation (6.1) shows the expected risk minimization.

$$R(\mathbf{w}^*) = min_{\mathbf{w}^* \in \mathbf{w}} E_{r, y \sim \mathscr{D}}(\ell(\mathbf{w}, r, y)). \tag{6.1}$$

Because the distribution $\mathscr{D}$ of the universe is unknown, we can only estimate the *empirical risk* $R_n(\mathbf{w})$ on training sample D. Equation (6.2) shows the empirical risk minimization.

$$R_n(\mathbf{w}) = min_{\mathbf{w}} \frac{1}{n} \sum_{i=1}^{n} \ell(h(\mathbf{w}, r_i), y_i) + \lambda\mu(\mathbf{w}), \tag{6.2}$$

where the regularizer $\lambda\mu(\mathbf{w})$ prevents the over-fitting.

By choosing different loss functions, ERM can be implemented to certain learning tasks such as linear regression [39], logistic regression [38], and kernel method [90, 105]. The private learning framework applied ERM to estimate the hypothesis set to choose an optimal output model $\mathbf{w}$. The utility of private ERM is measured by the difference between the real risk $R(\mathbf{w})$ and the private $R(\widehat{\mathbf{w}})$, which is defined as *risk bound*.

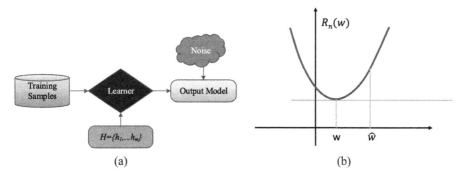

Fig. 6.4 (**a**) Output perturbation diagram. (**b**) Output perturbation

---

Algorithm 2 ERM with Output Perturbation

Require: Dataset D, and related parameters.
Ensure: output private model $\widehat{\mathbf{w}}$
   1. $\mathbf{w} = \arg\min_{\mathbf{w}} \frac{1}{n} \sum_i^n \ell(h(\mathbf{w}, r_i), y_i) + \lambda \mu(\mathbf{w})$
   2. Sample noise η from a Gamma distribution
   3. Compute $\widehat{\mathbf{w}} = \mathbf{w} + \eta$

6.2.2 Private Learning in ERM

When incorporating differential privacy into the current learning process, current works apply two methods via an ERM technique: *output perturbation* and *objective operation*. The *output perturbation* inserts noise into the output $\mathbf{w}$; while the *objective operation* adds noise to the objective function prior to learning. Based on both methods, we will discuss several learning algorithms, including *logistic regression* and *SVM*.

6.2.2.1 Output Perturbation

Chaudhuri et al. [38] proposed output perturbation solutions on *logistic regression*. Figure 6.4a illustrates the process and Fig. 6.4b shows that the $\widehat{\mathbf{w}}$ is obtained by moving original $\mathbf{w}$ on the horizontal axis. The sensitivity was $\frac{2}{n\lambda}$, which was associated with the regularizer parameter λ. With analysis, Chaudhuri et al. argued that the learning performance of this private algorithm will degrade with the decreasing of λ.

Their successive paper [39] formally analyzed the ERM framework. Algorithm 2 illustrates the output perturbation in a private ERM [39]. The author trained classifiers with different parameters on disjoint subsets of the data and utilized the exponential mechanism to privately choose parameters.

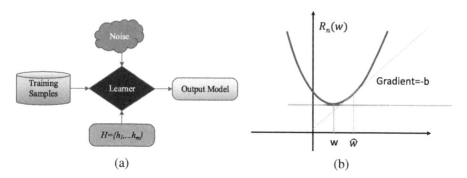

Fig. 6.5 (**a**) Objective perturbation diagram. (**b**) Objective perturbation

---

Algorithm 3 ERM with Objective Perturbation

Require: Dataset D, and related parameters.
Ensure: output private model $\widehat{\mathbf{w}}$
   1. Sample noise η from Gamma distribution;
   2. Compute $\mathbf{w} = \arg\min_{\mathbf{w}} \frac{1}{n} \sum_{i}^{n} \ell(h(\mathbf{w}, x_i), y_i) + \lambda r(\mathbf{w}) + <\eta, \mathbf{w}>$.

6.2.2.2 Objective Perturbation

Chaudhuri et al. [38, 39] also presented the objective perturbation solution by adding noise to the loss function $\ell(D)$, which is illustrated in Fig. 6.5. Particularly, Fig. 6.5a shows that the noise is directly added to the learner and Fig. 6.5b shows that the calibration of noise is associated with the gradient of $R_n(\mathbf{w})$. The formal ERM with objective perturbation is described in Algorithm 3 [39].

After analyzing the performance of both solutions, they conclude that when the function in the regression algorithm is convex and doubly differentiable, the objective perturbation outperforms the output perturbation. This is because regularization already changes the objective to protect against over-fitting, and changing the objective will not significantly impact performance. This claim is confirmed in terms of a risk bound in the next subsection.

6.2.2.3 Risk Bound in Different Learning Algorithms

Many machine learning algorithms can be implemented in privacy learning frameworks, including linear regression, online learning, deep learning, etc. With the exception of the learning process and the privacy budget, the risk bound is related to the dimension and the size of training samples. Bassily et al. [17] showed that the risk bound depends on $O(\sqrt{d}/n)$ under (ϵ, δ)-differential privacy and $O(d/n)$ under (ϵ)-differential privacy. These results show that the larger the dimension and the size, the lower utility that ERM can achieve. Table 6.5 lists several typical learning algorithms and their risk bounds.

Table 6.5 Private learning risk bound

Learning algorithm	References	Perturbation method	Loss function	Risk bound
Regression	[39]	Output/objective perturbation	Regularized MLE	Depends on $d \log d$/ depend on d
SVM	[105]	Output/objective perturbation	Arbitrary kernel, for example, polynomial kernels	Depends on $d^{1/3}/n^{2/3}$
	[196]	Output perturbation	Hinge-Loss function with invariant kernel, such as Gaussian kernel	Depends on $\sqrt{d}$
Online learning	[116]	Objective perturbation	Regularized MLE	Depends on $\sqrt{d}$ and time T
Deep learning	[5, 180]	Objective perturbation	Average of unmatched sample	N/A

Nearly all types of learning problems can be analyzed in terms of their risk bound. For example, most of the existing papers solve regression problems that have a single optimal objective. Ullman [222] considered the regression problem as a multiple objective optimization when the datasets are examined with different aims. Multiple convex minimization is considered as a set of queries that can be answered by a prevalent multiplicative weights method. Ullman implemented several single objective regression results in multiple objectives platform and their risk bounds are consistent with those original papers.

Kasiviswanathan et al. [116] considered private incremental regression in terms of streaming data. They combined continual release [69] with ERM technology to analyze the risk bound of several algorithms. Taking linear regression as an example, they continuously update a noisy version of the gradient function to minimize the MLE loss function. The risk bound depends on the $\sqrt{d}$ and the length of the stream: $O(\min\{\sqrt{d}, T\})$, where this bound is quite close to the normal linear regression private learner when the stream length T is large.

In addition, some papers consider private learning in a higher dimension dataset. Kifer et al. [128] gave the first results on private sparse regression with high dimensions. The authors designed the algorithm based on sub-sampling stability for support recovery using a LASSO estimator. Their following work [214] extended and improved the results with an algorithm based on a sample efficiency test of stability. Jain et al. [106] proposed an entropy regularized ERM using a sampling technique, This algorithm provides a risk bound that has a logarithmic dependence on d. Kasiviswanathan et al. [114] considered random projections in ERM frameworks. They provided a new private compress learning method with the risk bound related to the Gaussian width of the parameter space C in the random projection.

One of the most promising directions is the deep learning. Recent research focus has been devoted to the design of deep learning mechanisms. Abadi et al. [5] applied objective perturbation in a deep learning algorithm by defining the loss function as the penalty for mismatching training data. As the loss function is non-convex, they adapted a mini-batch stochastic gradient descent algorithm to solve the problem. The noise is added to every step of the stochastic gradient descent. In another pilot work [202], Shokri et al. designed a distributed deep learning model training system that enables multiple parties to jointly learn an accurate neural network. They implemented private stochastic gradient descent algorithms to achieve (ϵ, δ)-differential privacy within multiple parties. In the work of Phah et al. [180], the authors perturbed the objective functions of the traditional deep auto-encoder and designed the deep private auto-encoder algorithm by incorporating a Laplace mechanism.

6.2.2.4 Discussion

Private learning applies ERM to estimate the hypothesis set to select an optimal output **w**. It uses the output perturbation and the objective perturbation to ensure that the output satisfies differential privacy. A series of works have proven that this private learning is tractable for certain learning tasks, such as linear regression, SVM, and deep learning, etc.

Risk bounds are highly associated with the dimension of the dataset. Current research has decreased dependence on dimension to $O(d)$. Under certain assumptions, the dimension dependence could be further relaxed.

6.2.3 Sample Complexity Analysis

The second problem: *how many samples are needed in bounded accuracy?* is associated with sample complexity analysis. Sample complexity interprets the distance utility measurement in another way to show how many samples are needed to at least achieve a particular accuracy α. Probably approximately correct (PAC) learning [121] helps to measure the sample complexity of learning algorithms. Based on this theory, Blum et al. [29] and Kasiviswanathan et al. [115] proved that every finite concept class can be learned privately using a generic construction with a sample complexity of $O(VC(\mathscr{C}) \log |\mathscr{X}|)$ (we omit the other parameters). This is a higher sample complexity than a non-private learner who only needs a constant number of samples. Improving sample complexity is an essential issue.

The gap in sample complexity was studied in the follow-up papers and several methods were provided, which can be categorized into the following three groups:

- *the privacy requirement relaxation*;
- *the hypothesis relaxation*;
- and *semi-supervised learning*.

6.2.3.1 Relaxing Privacy Requirement

Beimel et al. [20] showed that when relaxing ϵ-differential privacy to (ϵ, δ)-differential privacy, sample complexity can be significantly decreased. Their follow-up paper decreases sample complexity to $O(\log(\sqrt{(d \cdot 1/\delta)}))$ [206].

Another way to relax the privacy requirement is to preserve privacy for the labels of samples, rather than entire attributes of samples. Chaudhuri et al. [37] assumed that except labels, attributes of samples are insensitive. They showed that any learning algorithm for a given hypothesis set that guarantees label privacy requires at least $\Omega(d')$ examples. Here, d' is the *doubling dimension* of the disagreement metric at a certain scale. The doubling dimension in $\mathcal{X}$ is the smallest positive integer d' such that every ball of $\mathcal{X}$ can be covered by $2^{d'}$ balls of half the radius.

The key idea of label privacy is to decrease the dimension of sensitivity information. It may give enough protection in a scenario where the content of the underlying samples is publicly known except their labels. In some other scenarios, however, attributes may be highly sensitive. For example, in the healthcare data, the identity (attributes) of the people should be protected as well as the diseases (label). Chaudhuri et al.'s relaxation is not applicable in this case.

6.2.3.2 Relaxing Hypothesis

This gap can also be closed by relaxing the requirement of the hypothesis. If the output hypothesis is selected from the learning concept $H \subseteq \mathcal{C}$, the learning process is defined as a proper learning. Otherwise, it is called an improper learning. For proper learning, the sample complexity is approximately $\Omega(d)$. If choosing improper learning, the sample complexity can be further decreased.

Beimel et al. [18] confirmed that when selecting a hypothesis that is not in $\mathcal{C}$, the sample complexity can be decreased to the constant. Their subsequent paper [19] proposed a probabilistic representation of $\mathcal{C}$ to improve the sample complexity. They considered a list of hypothesis collections $\{H_1, ..., H_m\}$ rather than just one collection of H to represent $\mathcal{C}$. The authors assumed that, when sampling H_i from the hypothesis list, there will be $h \in H$ close to c in high probability. The sample complexity can be reduced to $O(\max(\ln |H_i|))$.

This improvement in sample complexity comes at the cost of an increased workload on evaluation, however. The learner will have to exponentially evaluate many points that are far from the concept set $\mathcal{C}$. In general, for a private learner, if $H = \mathcal{C}$, the sample complexity is $O(d)$ and the time for evaluation is constant. If $H \neq \mathcal{C}$, there is constant sample complexity but $O(\exp(d))$ time for evaluation.

6.2.3.3 Semi-Supervised Learning

Semi-supervised learning is a useful method for reducing the complexity of labeled samples. Beimel et al. [21] proposed a private learner by introducing semi-supervised learning to active learnings. The method starts with an unlabeled dataset to create a synthetic dataset for a class $\mathscr{C}$. This synthetic dataset is then used to choose a subset of the hypotheses with a size of $2^{O(VC(\mathscr{C}))}$. In the last step the authors apply $O(VC(\mathscr{C}))$ labeled examples to choose the target synthetic dataset according to the hypotheses set.

In this process, the sample complexity of the labeled samples is $O(\frac{VC(\mathscr{C})}{\alpha^3 \epsilon})$ while for the unlabeled samples it is $O(\frac{d \cdot VC(\mathscr{C})}{\alpha^3 \epsilon})$. Comparing the sample complexity of labeled and unlabeled samples, this private learner uses a constant number of labeled samples and $O(d)$ unlabeled samples.

6.2.3.4 Discussion

Table 6.6 compares different methods in terms of their sample complexity. To make the results clear, we utilize the *VC dimension* representation and omit other parameters such as α, β and ϵ.

6.2.4 Summary on Private Learning Framework

The private learning framework focuses on privacy preserving in the learning problem. Its foundation theory is the ERM and PAC learning theory. The more samples an algorithm can obtain, the more accurate will be the result. After applied the PAC learning theory in privacy learning, the line of research works proposed

Table 6.6 Comparison of the sample complexity of different methods

Method	References	Description	Sample complexity	Privacy level
Original	Kasiviswanathan et al. [115], Blum et al. [29]	Uses an exponential mechanism to search h	$O(VC(\mathscr{C})log|\mathscr{X}|)$	ϵ
Relaxing privacy level	Beimel et al. [20]	From ϵ to ϵ, δ	$O(\log(1/\delta))$	(ϵ, δ)
	Steinke et al. [206]	From ϵ to ϵ, δ	$O(\log(\sqrt{d} \cdot 1/\delta))$	(ϵ, δ)
	Chaudhuri et al. [37]	Only preserve privacy for labels	$\Omega(d')$ (d' was the adjusted dimension which is lower than d)	(ϵ)
Relaxing hypothesis	Beimel et al. [18, 19]	$H \neq \mathscr{C}$ and set a group of H to privately select a h.	$O(\max(ln|H_i|))$	ϵ
Semi-supervised learning	Beimel [21]	Use labeled data to search h	$O(dVC(\mathscr{C}))$ (labeled) $O(VC(\mathscr{C}))$ (unlabeled)	ϵ

accuracy bounds on the private learner, and tried to decrease the sample complexity in a fixed accuracy bound. Current research narrows down the sample complexity gap between the private and non-private learning processes, which means that we can learn privately by using acceptable number of samples.

The private learning framework is only concerned with supervised learning algorithms; in addition, these algorithms should be PAC learnable. These strict constraints hinder the development of the privacy learning framework, making it currently actively developing in theory, but still impractical for real applications.

6.3   Summary of Differentially Private Data Analysis

As the most prevalent framework, the Laplace/exponential framework has been widely used. The most prominent advantages are its flexibility and simplicity; it can freely introduce Laplace and exponential mechanisms into various types of algorithms. For the non-experts on privacy, it proposed a possible way to ensure that the results satisfy the requirement of differential privacy. However, the essential challenge for this framework is accuracy of the results, especially for those algorithms whose operations have high sensitivities. They lead to a large volume of noise in the analysis results. The Laplace/exponential framework is widely used in current applications but there is still room for further improvement.

Private learning frameworks combine differential privacy with diverse learning algorithms to preserve the privacy of the learning samples. The foundation theories are ERM and PAC learning. ERM helps to select the optimal learning model by transferring the learning process into a convex minimization problem. Differential privacy either adds noise to the output models or to the convex objective functions. PAC learning estimates the relationship between the number of learning samples and the model's accuracy. The more samples an algorithm can obtain, the more accurate the result will be. As privacy learning results in higher sample complexity, researchers are currently trying to narrow the sample complexity gap between private and non-private learning processes so that private learning is feasible with an acceptable number of samples.

Private learning frameworks, however, have some constraints. ERM requires that the objective function should be convex and L-Lipschitz. PAC learning can only be applied when the algorithm is PAC learnable. These constraints hinder the practical development of privacy learning frameworks and make them an actively developing theoretical proposition, yet still impractical for real applications.

Chapter 7
Differentially Private Deep Learning

7.1 Introduction

In recent years, deep learning has rapidly become one of the most successful approaches to machine learning. Deep learning takes advantage of the increasing amount of available computation to process big data. In addition, the new algorithms and architectures being developed for deep neural networks are accelerating the progress of various fields, including image classification, speech recognition, natural language understanding, social networks analysis, bioinformatics, and language translation [132].

The essential idea of deep learning is to apply a multiple-layer structure to extract complex features from high-dimensional data and use those features to build models. The multiple-layer structure consists of neurons. The neurons in each layer receive a finite number of output neurons from the previous layer along with their associated weights. The aim of the training process is to adjust the weights of these neurons to fit the training samples. In practice, a stochastic gradient descent (SGD) procedure is one of the most popular ways to achieve this goal.

Like other machine learning models, deep learning models are susceptible to several types of attacks. For example, a centralized collection of photos, speech, and video clips from millions of individuals might meet with privacy risks when shared with others [202]. Learning models can also disclose sensitive information. Fredrikson et al. proposed a model-inversion attack that recovers images from a facial recognition system [81]. Adabi et al. [6] assumed that the adversaries would not only have access to the trained model but may also have the full knowledge of the training mechanism and the model parameters. Phah et al. [180] considered a general adversarial setting in which potential privacy leaks can stem from malicious inference with the model's inputs and outputs.

Differential privacy can be integrated with deep learning to tackle these privacy issues. However, directly applying Laplace noise within a deep learning model yields inferior performance for several reasons:

© Springer International Publishing AG 2017
T. Zhu et al., *Differential Privacy and Applications*,
Advances in Information Security 69, DOI 10.1007/978-3-319-62004-6_7

Table 7.1 Differentially private deep learning methods

Related work	Adversarial setting	System setting	Privacy guarantee method
Shokri et al. [202]	Additional capabilities	Distributed system	Differentially private SGD algorithm with convex objective functions
Adabi et al. [6]	Additional capabilities	Centralized system	Differentially private SGD algorithm with non-convex objective functions
Phah et al. [180]	General capabilities	Centralized system	Objective function perturbation of deep auto-encoder

- High sensitivity: in deep learning training processes, the sensitivity of both the objective function and the model's output during perturbation is quite high.
- Limited privacy budget: iterative training processes divide the privacy budget into several pieces, which leads to high levels of noise in the final result.

Several possible solutions to these challenges have been explored. For example, Adabi et al. [6] clipped the objective function to bound its sensitivity and applied a *moment accountant* method to the objective function to form an optimal privacy composition. Phah et al. [180] applied a functional mechanism to perturb the objective function and decrease noise. In general, the current research on differentially private deep learning can be classified according to three criteria, as shown in Table 7.1:

Its adversarial setting.   In the work of Shokri et al. [202] and Adabi et al. [6], assumed that adversaries would not only have access to the trained model but would also have full knowledge of the training mechanism and the model's parameters. Whereas, Phah et al. [180] considered a more general adversarial setting, where potential privacy leaks might arise from malicious inference with a model's inputs and outputs.

Its system setting.   In a pilot study [202], Shokri et al. designed a distributed deep learning model training system that enables multiple parities to jointly learn an accurate neural network. However, both Adabi et al. [6] and Phah et al. [180] considered a centralized system setting where data are held centrally.

The privacy guarantee method used.   The methods for guaranteeing differential privacy can be classified into two types. The first type adds noise to the execution process of the optimization algorithm. The second perturbs the objective by adding differentially private noise to the objective functions before the learning procedure Shokri et al. [202] and Adabi et al. [6] designed a differentially private SGD algorithm by introducing a sparse vector technique [73], while Adabi et al.'s [6] SGD approach relied on a Gaussian mechanism. Phah et al. [180], perturbed the objective functions of a conventional deep auto-encoder and designed a deep private auto-encoder algorithm that incorporates a functional mechanism.

7.2 Preliminary

Based on artificial neural networks and the rapid development of cloud computing and big data techniques, deep learning aims to extract nonlinear features and functions from massive data to train complex models and their numerous parameters. The main difference between deep learning and traditional machine learning is that the former involves learning feature representation, while the latter always leverages hand-designed features. Thus, the two main challenges in deep learning are: how to automatically learn the values of the parameters, or weights, of each neuron from the training data; and how to minimize the objective functions of the neural network.

7.2.1 Deep Learning Structure

Suppose $D = \{(x_1, y_1), \ldots, (x_n, y_n)\}$ is a set of samples (the training dataset) drawn from a universe $\mathscr{X}$, where $y_i \in \{0, 1\}$ is the label for each sample x_i, The layer contains the samples that we can observe, the link between layers are and a multiple-layer network is the most popular structure for deep learning architectures. The network includes an input layer, an output layer, and some hidden layers. Figure 7.1 is a typical multiple-layer network with three hidden layers. The bottom layer is the input layer; it accepts the training samples and can be observed. The three middle layers are hidden layers that extract increasingly abstract features from the input samples. The output layer contains the final results from processing the samples,

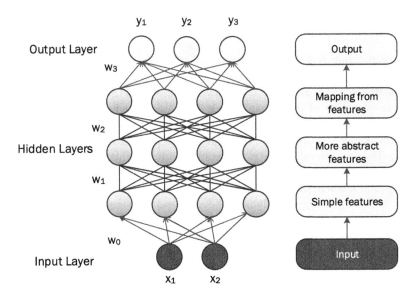

Fig. 7.1 Deep learning structure

Fig. 7.2 Neuron input and
output

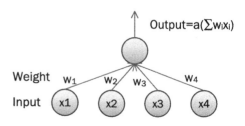

such as classification labels. This process can be interpreted as the right-hand side
of the flowchart shown in Fig. 7.1, which provides a high-level schematic of how
each layer works.

The circles in each layer represent neurons. Each neuron receives a finite number
of outputs from the neurons in the previous layer along with its associated weight.
A neuron's output is calculated by applying a nonlinear activation function to all the
input values. Figure 7.2 shows a neuron's inputs and its output. Suppose the inputs
of a neuron are $\mathbf{x} = (x_1, x_2, x_3, x_4)$ and the weights are $\mathbf{w} = (w_1, w_2, w_3, w_4)$, the
output of the neuron is $a(\sum w_i \cdot x_i)$, where $a()$ is the activation function.

The activation function is normally a nonlinear function that transforms the
weighted sum of the inputs. The most popular activation functions are sigmoid,
tanh or a rectified linear function.

- Sigmoid function: $sigmod(x) = \frac{1}{1+e^{-x}}$
- Tanh function: $tanh(x) = \frac{2}{1+e^{-2x}} - 1$
- Rectified linear function:

$$rec(x) = \begin{cases} 0 & \text{for } x < 0 \\ x & \text{for } x \geq 0. \end{cases}$$

The tanh function is a re-scaled version of the sigmoid function with an output
range of $[-1, 1]$ instead of $[0, 1]$. The rectified linear function is a piece-wise linear
function that saturates at exactly 0 whenever the input x is less than 0.

In the designing of the multiple-layer structures, weights $\mathbf{w}$ are derived from
an objective function $J(\mathbf{w}, \mathbf{x})$. Finding $\mathbf{w}$ is an optimization process that yields an
acceptably small loss of $J(\mathbf{w}, \mathbf{x})$. There are several possible definitions for function
$J(\mathbf{w}, \mathbf{x})$. The most prevalent one is the average of the loss over the training samples
$\{x_1, x_2, \ldots x_n\}$, where $J(\mathbf{w}) = \frac{1}{n} \sum_i J(\mathbf{w}, x_i)$. As $J(\mathbf{w}, \mathbf{x})$ is normally a non-convex
function, a gradient descent method is applied to estimate $\mathbf{w}$.

The following sections briefly describe SGD and deep auto-encoders-two representative concepts in this field. SGD, presented in Sect. 7.2.2, plays an important
role in most deep learning optimization algorithms. Deep auto-encoders are a fundamental part of deep learning model structures and are presented in Sect. 7.2.2.1.

7.2.2 Stochastic Gradient Descent

In a typical deep learning structure, there may be hundreds of millions of adjustable weights. The weight learning process can be considered as a nonlinear optimization problem, while the objective functions measure the errors. One of the most popular algorithms for properly adjusting the weight vectors is gradient descent. In this approach, a gradient vector is used to indicate the amount an error would increase or decrease if a given weight was increased by a tiny amount, and the weight vector for that weight is then adjusted in the opposite direction [132].

However, when applied to learning tasks on a large dataset, gradient descent algorithms are far from efficient. SGD is an extension to gradient descent, and these algorithms can significantly reduce computational costs compared to their predecessors. SGD algorithms have therefore been widely used in deep learning.

Consider an input layer that contains only a few training samples (x_i, y_i). SGD computes the average gradient for those examples and adjusts the weights iteratively. The process is repeated for many small sets of examples from the training set until the average of the objective function stops decreasing. Due to the difficulty of minimizing the loss function $J(\mathbf{w}, \mathbf{x})$ in complex networks, a batch of training data are randomly selected to form a subset S (mini-batch) at each step of the SGD algorithm. Hence, the gradient $\nabla_{\mathbf{w}} J(\mathbf{w}, \mathbf{x})$ can be estimated as:

Next, the weights $\mathbf{w}$ of the objective $J(\mathbf{w})$ are updated as follows:

$$\mathbf{w}_{t+1} = \mathbf{w}_t - \alpha_t g_S, g_S = \frac{1}{|S|} \sum_{x \in S} \nabla_w J(\mathbf{w}, \mathbf{x}), \tag{7.1}$$

where α is the learning rate and t is the current iteration round.

Algorithm 1 shows the pseudo-code for the SGD algorithm. Note that a *Back-Propagation* algorithm is always used to compute the derivative of the loss function $J(\mathbf{w}, \mathbf{x})$ with respect to each parameter (lines 4–6). In practice, the repeating statements (lines 2–9) are are only executed a fixed number of times T to save overall time costs.

Algorithm 1 SGD Algorithm

Require: Training dataset $\mathbf{x} = \{x_1, x_2, \ldots, x_n\}$, loss function $J(\mathbf{w}, \mathbf{x}) = \frac{1}{n} \sum_{i=1} J(\mathbf{w}, x_i)$, learning rate α.

Ensure: $\mathbf{w}$.

1: Initialize $\mathbf{w}_0$ randomly;
2: **repeat**
3:     Randomly take samples S_t from the training dataset $\mathbf{x}$;
4:     **for** each $i \in S_t$ **do**
5:         compute $g_t(x_i) \leftarrow \nabla_{\mathbf{w}_t} J(\mathbf{w}_t, x_i)$;   {**Compute gradient**}
6:     **end for**
7:     compute $g_t \leftarrow \frac{1}{|S_t|} \sum_{i \in S_t} g_t(x_i)$;
8:     update $\mathbf{w}_{t+1} \leftarrow \mathbf{w}_t - \alpha g_t$;   {**Descent**}
9: **until** an approximate minimum is obtained.
10: **return** $\mathbf{w}$.

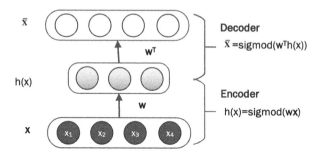

Fig. 7.3 Auto-encoder

7.2.2.1 Deep Auto-Encoder

As a basic deep learning model, the *Deep Auto-Encoders* [22] is composed of multiple *Auto-Encoders*. Generally each auto-encoder has two parts: an encoder that encodes the input into a representation using the function $h(\mathbf{x})$, and a decoder that reconstructs the input from the encoded representation using the function $\bar{x} = d(h)$. The auto-encoders aim to minimize the negative log-likelihood of the reconstructions.

Figure 7.3 shows the structure of an auto-encoder. It has a layer of input neurons, a hidden layer in the middle, and an output layer at the end. Generating the hidden layer is considered to be the encoding process, during which a sigmod function is used to activate the neurons in the hidden layer. Equation (7.2) shows the creation of $h(\mathbf{x})$

$$h(\mathbf{x}) = sigmod(\mathbf{wx}). \tag{7.2}$$

From the hidden layer to the output layer, the decoder uses the weight vector w:

$$\bar{\mathbf{x}} = sigmod(h(\mathbf{w}^T\mathbf{x})). \tag{7.3}$$

The auto-encoder model is trained to minimize the difference between the input and the output. There are two types of loss functions. For a binary input $\mathbf{x}$, the auto-encoder minimizes the negative log-likelihood of the reconstruction as shown in Eq. (7.4)

$$J(\mathbf{w}, \mathbf{x}) = -logPr(\mathbf{x}|\bar{\mathbf{x}}, \mathbf{w}) = -\sum_{n}^{i=1}(x_i log\bar{x}_i + (1 - x_i)log(1 - \bar{x}_i)). \tag{7.4}$$

This loss function actually measures the cross-entropy error of a binomial problem. When the inputs are real values, this loss function normally measures the sum of the squared Euclidean distance between the output and input values. as shown in Eq. (7.5).

$$J(\mathbf{w}, \mathbf{x}) = \frac{1}{2} \sum_i (\bar{x}_i - x_i)^2. \tag{7.5}$$

Gradient methods can also be used to minimize loss functions.

Multiple auto-encoders can be stacked to form a deep auto-encoder. Or, some-times an output layer can be added to the top of a deep auto-encoder to predict Y using a single binomial variable. In these cases, the cross-entropy error serves as the loss function to measure the difference between y_i and y:

$$E(Y_T, \theta) = -\sum_{i=1}^{|Y_T|} [y_i \log \widehat{y}_i + (1 - y_i) \log(1 - \widehat{y}_i)], \tag{7.6}$$

where Y_T is the set of labeled data used to train the model.

7.3   Differentially Private Deep Learning

The common interests and permission requirements of multiple parties demand that the privacy preserving goals of deep learning include the protection of both the training datasets and the training parameters. The main methods for guaranteeing differential privacy in deep learning models are: (1) adding noise to the execution process of an existing optimization algorithm; or (2) perturbing the objective functions of the given optimization problem. A differentially private SGD algorithm is representative of the former method, while the latter method typically uses a deep private auto-encoder. This section briefly reviews the rationales for these related works.

One straightforward way to protect the privacy of training data against leaks is to simply add noise to the trained parameters resulting from the learning algorithm. That is, the privacy-enhancing techniques are not applied to the internal training process but rather to its outputs. However, due to the difficulties in fine-grained noise analysis when adopting this approach, adding too much noise would further distort the utility of the learning algorithm. Motivated by this problem, the extant literature has presented several sophisticated approaches that aim to guarantee the differential privacy of training data during the training process. As an essential component of the training process, the SGD algorithm has also been enhanced in various differentially private techniques. Following Sections first presents the basic Laplace method, and then presents private SGD, private auto-encoder and distributed private SGD.

Algorithm 2 Basic Laplace Method

Require: $D = \{(x_1, y_1), \ldots, (x_n, y_n)\}$, $J(\mathbf{w}, \mathbf{x})$, learning rate α, noise scale σ, batch size S, privacy
   budget ϵ.
Ensure: w.
   1. initialize $\mathbf{w}_0$ randomly;
   2. $\epsilon_t = \epsilon/T$;
   for $t = 0, \ldots, T - 1$ **do**
      3. take a random sample set S_t from D;
      4. compute gradient: for each $i \in S_t$, compute $g_t(x_i) = \nabla_{w_t} J(w_t, x_i)$;
      5. add noise: $\widehat{g}_t = \frac{1}{S}(\sum_i g_t(x_i) + Lap(\Delta J/\epsilon_t))$;
      6. descent: $\widehat{\mathbf{w}}_{t+1} = \widehat{w}_t - \alpha_t \widehat{g}_t$;
   end for
   7. $\widehat{w} = \widehat{\mathbf{w}}_T$.

7.3.1 Basic Laplace Method

The most direct method is to add noise to the SGD process. Laplace noise can be
added to each iterative gradient descent iterative round t, as shown in Eq. (7.7)

$$\widehat{g}(x_i) = \nabla_{\mathbf{w}_t} J(\mathbf{w}_t, x_i) + Lap(\Delta J/\epsilon_t). \tag{7.7}$$

Then the weight is then estimated by $\widehat{g}$ through Eq. (7.8):

$$\widehat{\mathbf{w}}_{t+1} = \widehat{\mathbf{w}}_t - \alpha_t \nabla_{\mathbf{w}} \widehat{g}_t(\mathbf{w}; x_i). \tag{7.8}$$

Algorithm 2 shows the procedure for a basic Laplace method to train a model
with weight set w by minimizing the empirical objective function $J(\mathbf{w}, \mathbf{x})$. The
weight set is initialized by a set of random numbers in Step 1. Suppose there are a
total of T rounds in this iterative gradient descent model, the privacy budget would
be divided into T pieces in Step 2. Step 3 takes a random batch sample set S_t from
the training set. For a single sample $x_i \in S_t$, partial differential $g_t(x_i)$ of the function
$J(\mathbf{w}_t, x_i)$ is computed in Step 4. Step 5 adds Laplace noise to the $g_t(x_i)$ based on the
sensitivity of $J(\mathbf{w_t}, x_i)$ and the privacy budget ϵ_t. In the descent Step 6, the weights
are estimated from the weights in the last round and the noisy function $\widehat{g}(x_i)$. After
all rounds are finalized, weighs $\mathbf{w}$ is determined by the weights of in the last round.

Clearly, Step 5 guarantees that the output result $\widehat{\mathbf{w}}$ satisfies ϵ-differential privacy.
However, as mentioned in Sect. 7.1, this basic Laplace method produces an inferior
learning model based on $\widehat{\mathbf{w}}$. The noise $Lap(\Delta J/\epsilon_t)$ is quite large as there is no a
priori bound on the size of the gradients, which leads to a high ΔJ. In addition,
due to the vast number of iterative rounds in the gradient process, ϵ_t is quite small,
which lead to a high level of noise. If there were 1000 iterations and the total privacy
budget for each sample was set to 1, each iteration would only be allocated 0.001 of
the privacy budget. Therefore, applying the basic Laplace method to deep learning
models is impractical in real-world cases. However, researchers have proposed
various methods to improve the quality of these models.

Algorithm 3 Differentially Private SGD

Require: $D = \{(x_1, y_1), \ldots, (x_n, y_n)\}$, $J(w)$, learning rate α, noise scale σ, group size L, gradient norm bound C.

Ensure: w.

   1. initialize w_0 randomly;

  for $t = 0, \ldots, T - 1$ **do**

      2. take a random sample L_t with probability L/n;

      3. compute gradient: for each $i \in L_t$, compute $g_t(x_i) = \nabla_{w_t} J(w_t, (x_i, y_i))$;

      4. norm clip gradient: $\bar{g}_t = g_t(x_i) / \max(1, \frac{\|g_t(x_i)\|_2}{C})$;

      5. add noise: $\widehat{g}_t = \frac{1}{L}(\sum_i \bar{g}_t(x_i) + \mathcal{N}(0, \sigma^2 C^2 I))$;

      6. descent: $\widehat{w}_{t+1} = \widehat{w}_t - \alpha_t \widehat{g}_t$;

  end for

   7. $\widehat{w} = \widehat{w}_T$;

   8. compute the overall privacy loss (ϵ, δ) using a privacy accounting method.

7.3.2   Private SGD Method

Adabi et al. [6] extended the basic Laplace method and proposed a private SGD algorithm. Similar to the basic Laplace mechanism, the noise is added to the objective function $J(\mathbf{w}, \mathbf{x})$. However, three additional technologies improve the performance of the output model.

- Norm clipping of the objective function J to bound its sensitivity.
- Grouping several batches together then adding noise to the group.
- Applying moment accountant to the objective functions to form an optimal privacy composition

A brief outline of their algorithm is presented before analyzing the above three technologies in detail.

Algorithm 3 presents the differentially private SGD algorithm. It has some differences to Algorithm 2. First, Algorithm 3 includes several new parameters, including noise scale σ, group size L, and gradient norm bound C. These parameters are used when adding noise, grouping batches and in norm clipping, respectively. Similar to the basic Laplace method, the private SGD method initializes $\mathbf{w}_0$ with random values in Step 1. However, in Step 2, instead of using a batch of samples S_t, a larger sample group L_t is sampled. This can be considered as a merger of several batch samples. The aim of Step 2 is to decrease the total noise added to the private SGD method. The gradient computing in Step 3 is similar to the basic Laplace method. In Step 4, a private SGD method clips the gradient with a constant C, aiming to bound the sensitivity of g_t. Step 5 adds calibrated Gaussian noise to g_t and calculates the weights using a noisy version of g_t in Step 6. Finally, in addition to the output of weights in Step 7, the privacy loss of the method is derived based on a privacy accounting technique in Step 8.

7.3.2.1 Norm Clipping

One of the challenges in differentially private deep learning is the high sensitivity of the objective function. To bound the influence of each individual sample on $g(\mathbf{x})$, which is the gradient of $J(\mathbf{w}, \mathbf{x})$, the private SGD method clips each gradient in ℓ_2 norm by using a predefined threshold for C, as shown in Eq. (7.9).

$$\|g(\mathbf{x})\|_2 = \begin{cases} \|g(\mathbf{x})\|_2 & \text{for } \|g(\mathbf{x})\|_2 \leq C \\ C & \text{for } \|g(x)\|_2 > C. \end{cases} \tag{7.9}$$

In the norm clipping step, the gradient vector $g_t(x_i)$ in round t will be replaced by $g_t(x_i)/\max(1, \frac{\|g_t(x_i)\|_2}{C})$, and the sensitivity of the $g_t(x_i)$ is bounded to C: $\Delta_2 g = C$. It worth noting that in a multi-layer neural network, each layer can be set to a different clipping threshold C.

7.3.2.2 Grouping Batches

As SGD normally estimates the gradient of $J(\mathbf{w}, \mathbf{x})$ by computing the gradient of the loss on a batch of samples and taking the average, noise has to be added to each batch. Hence, a smaller batch size leads to more noise in total. To decrease the amount of added noise, the private SGD method groups several batches together into a larger group L, called a *lot*, and adds the noise to the lot. In practice, each lot is created independently by picking each sample with a probability of L/n.

7.3.2.3 Privacy Composition

The proposed private SGD method applies a Gaussian mechanism with a sensitivity equal to C. According to the Gaussian mechanism defined in Chap. 2, each *lot* in private SGD preserves $(T\epsilon, T\delta)$-differential privacy when $\sigma = \Delta_2 f \sqrt{2\ln(2/\delta)}/\epsilon$. As there are L/n lots in the dataset, the final output preserves $(TL/n\epsilon, TL/n\delta)$-differential privacy. This forms loose bound for the output in terms of sequential composition.

However, to improve the bound of the privacy loss, the private SGD method introduces a stronger accounting method, moments accountant, which improves the bound to $O(\sqrt{T}L/n\epsilon, \delta)$-differential privacy for an appropriately chosen noise scale and a threshold of C.

Theorem 7.1 *There exist the constants c_1 and c_2. Given the sampling probability L/n, and the number of iterative round T, for any $\epsilon < c_1 TL/n$, the private SGD is (ϵ, δ)-differential privacy for any $\delta > 0$ if we choose*

$$\sigma \geq c_2 \frac{L/n\sqrt{T\log(1/\delta)}}{\epsilon}. \tag{7.10}$$

Algorithm 4 Deep Private Auto-Encoder Algorithm

Require: $\mathbf{x} = \{(x_1, \ldots, x_n\}, J(\mathbf{w}, \mathbf{x})$
Ensure: $\widehat{\theta}$.
    1. Derive a polynomial approximation of $J(\mathbf{w}, \mathbf{x})$, denoted as $\tilde{J}(\mathbf{w}, \mathbf{x})$
    2. Perturb the function $\tilde{J}(\mathbf{w}, \mathbf{x})$ to generate perturbed function of $\widehat{J}(\mathbf{w}, \mathbf{x})$ by using a functional mechanism.
    3. Compute $\widehat{\mathbf{w}} = \arg\min \widehat{J}(\mathbf{w}, \mathbf{x})$;
    4. Stack k private auto-encoders;
    5. Derive and perturb the polynomial approximation of the cross-entropy error $E(\theta)$;
    6. Compute $\boldsymbol{\theta} : \widehat{\theta} = argminE(\theta)$.

7.3.3 Deep Private Auto-Encoder Method

Phah et al. [180] perturbed the objective functions of a traditional deep auto-encoder and designed a deep private auto-encoder algorithm that incorporates a Laplace mechanism.

7.3.3.1 Deep Private Auto-Encoder Algorithm

Phah et al. [180] deep private auto-encoder algorithm perturbs the loss function $J(\mathbf{w}, \mathbf{x})$. But, unlike the basic Laplace or the previously mentioned private SGD method, this deep private auto-encoder algorithm applies a functional mechanism [252] to a perturbed $J(\mathbf{w}, \mathbf{x})$. This proposed deep private auto-encoder model is used to make binomial prediction.

Algorithm 4 shows the key steps of the deep private auto-encoder algorithm. The first step applies a Taylor expansion to approximate the loss function $J(\mathbf{w}, \mathbf{x})$, and then the Taylor expansion is perturbed using a functional mechanism in Step 2. After creating the perturbed loss function $\widehat{J}(\mathbf{w}, \mathbf{x})$, Step 3 computes the weight vector using gradient descent. Step 4 stacks private auto-encoders to construct a deep private auto-encoders. Here, the previous hidden layer can be considered as the input of the next auto-encoder. The first four steps aim to create a deep private auto-encoder model. Step 5 focuses on a binomial prediction problem in the top layer, which includes a single binomial variable to predict y and apply a loss function to the cross-entropy error $E(\theta)$. The function $E(\theta)$ is also perturbed by a functional mechanism. The final step computes $\widehat{\theta}$ by minimizing $\widehat{E}(\theta)$.

Figure 7.4 shows the resulting structure of the deep private auto-encoder model. It is worth noting that, before stacking multiple hidden layers, a normalization layer, denoted as η, is inserted after each hidden layer. The normalization layer guarantees all data assumptions and that differential privacy will be satisfied when training the next hidden layer.

Fig. 7.4 Structure of the deep private auto-encoder

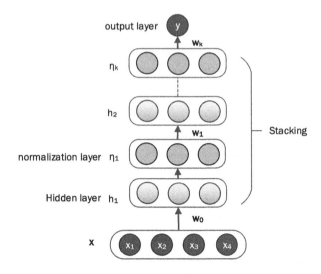

7.3.3.2   Functional Mechanism

As proposed by Zhang et al. [252], the functional mechanism aims to perturb the objective functions by adding Laplace noise into every coefficient λ_ϕ of this function's approximate polynomial form. In the deep private auto-encoder method, before adopting a functional mechanism to perturb the data reconstruction function $J(\mathbf{w}, \mathbf{x})$, the approximate polynomials from Step 2 and the *Cross-Entropy Error* $E(\theta)$ (Step 5) are first derived by leveraging the Taylor expansion (Step 1 and Step 5, respectively).

7.3.3.3   Sensitivity Measurements

The deep private auto-encoder algorithm contains two perturbations. The first perturbation is the loss function $J(\mathbf{w}, \mathbf{x})$ during the training process; the second is the loss function $E(\theta)$ in the prediction process. As both perturbations apply a Laplace mechanism, each uses a different sensitivity, as shown in Eqs. (7.11) and (7.12), respectively, are used in the perturbation.

$$\triangle J = d\left(b + \frac{1}{4}b^2\right), \tag{7.11}$$

where d is the dimension of samples and b is the number of hidden variables.

$$\triangle E = \left(|\eta| + \frac{1}{4}|\eta|^2\right), \tag{7.12}$$

where $|\eta|$ is the number of variables in the normalization layer. Both sensitivities are highly related to the variables in the hidden layer, or normalization layer. Increasing the variables in the hidden layer may increase the prediction accuracy in the learning process; however, this may also increase the amount of noise added to the loss functions.

7.3.4   Distributed Private SGD

In addition to the potential privacy leaks yielded by non-private algorithms, the basic architecture of centralized deep learning systems can also cause several privacy issues, especially when considering the concerns of everyone involved. For example, once an individual user has contributed information to a massive dataset that is subsequently used to develop a complex learning model, it is nearly impossible for that person to control or influence what happens to their data. It is relatively hard for users to completely trust companies in the big data era. Further, when considering the co-operation required between multiple companies and institutes aiming to construct models using joint data, achieving proprietary data privacy for both companies while jointly sharing the data used greatly compounds these difficulties. Distributed private deep learning architectures are in need of development. However, SGD algorithms can be parallelized and executed asynchronously [55, 191, 261], These advantages led Shokris [202] to propose a distributed selective SGD (DSSGD) protocol, in which selective parameter sharing was the key novelty.

At the outset, all the participants involved in the DSSGD agree on a common learning objective. A parameter server is used to maintain the settings for all the parameters $\omega^{(global)}$ and their latest values.

Each participant i downloads the parameters $\mathbf{w}^{(i)}$, replacing their local parameters with the latest values from the server. Each participant then executes the SGD algorithm and uploads a portion of the computation results $\nabla \mathbf{w}_P^{(i)}$ to the parameter server. The server collates the updates from multiple participants, maintains the latest parameter values, and makes the results available to all participants. Pseudo-code for the participant's side of the DSSGD protocol is shown below. Some of the notations from the original description [202] have been replaced to maintain consistency with the previous contents of this chapter. Note that the parameters participant i uploads to the server are selected according to one of the following criteria:

- *largest values:* sort gradients in $\nabla \mathbf{w}^{(i)}$ and upload β_u fraction of them, starting with the largest.
- *random with threshold:* randomly subsample the gradients with values above threshold τ.

These selection criteria are fixed for the entire training set.

Algorithm 5 DSSGD for Participant i

Require: The initial parameters $\mathbf{w}^{(i)}$ that are chosen by the server, learning rate α, fraction of parameters selected for download β_d, fraction of parameters selected for upload β_u.

1: **repeat**
2:　　Download $\beta_d \times |\mathbf{w}^{(i)}|$ parameters from server and replace the corresponding local parameters;
3:　　Run SGD on the local dataset and update the local parameters $\mathbf{w}^{(i)}$;
4:　　Compute gradient vector $\nabla \mathbf{w}^{(i)}$ which is the vector of changes in all local parameters due to SGD;
5:　　Upload $\nabla \mathbf{w}_P^{(i)}$ to the parameter server, where P is the set of indices of at most $\beta_u \times |\mathbf{w}^{(i)}|$ gradients that are selected according to some criteria;
6: **until** an approximate minimum is obtained.

7.3.4.1　Sparse Vector Technique

The above DSSGD protocol is only able to prevent direct privacy leaks that occur as a result of sharing a participant's local training data with others. Therefore, Shokri et al. [202] also adopted a sparse vector technique (SVT) [73] to prevent indirect privacy leaks caused by participants sharing locally updated parameters. SVTs were originally proposed as a way of answering some queries in interactive settings without the need to consume any of the privacy budget. This method is based on the intuition that, in some situations, only the queries with answers above a certain threshold need to be considered; those below the threshold do not. In this way, SVTs first compare a noisy query answer and a noisy threshold, then release an output vector that indicates whether the answer is above or below the given threshold. Algorithm 6 shows the pseudo-code for a differentially private DSSGD for participant i using an SVT.

Algorithm 6 Differentially Private DSSGD for Participant i Using an SVT

Require: The total privacy budget ϵ, the sensitivity Δf of each gradient, the maximum number $c = \beta_u \times |\mathbf{w}^{(i)}|$ of uploaded gradients, the bound γ on gradient values shared with other participants, the threshold τ for gradient selection.

1: Initialize $\epsilon_1 = \frac{8}{9}\epsilon$, generate fresh random noise $\rho = \mathsf{Lap}(2c\Delta f/\epsilon_1)$;
2: Initialize $\epsilon_2 = \frac{1}{9}\epsilon$, count = 0;
3: Randomly select a gradient $\nabla w_j^{(i)}$;
4: Generate fresh random noise $v_j = \mathsf{Lap}(4c\Delta f/\epsilon_1)$;
5: **if** abs(bound$(\nabla w_j^{(i)}, \gamma)) + v_j \geq \tau + \rho$ **then**
6:　　Generate fresh random noise $v_j' = \mathsf{Lap}(2c\Delta f/\epsilon_2)$;
7:　　uploaded bound$(\nabla w_j^{(i)} + v_j', \gamma)$ to the parameter server;
8:　　charge $\frac{\epsilon}{c}$ to the privacy budget;
9:　　count = count + 1, if count $\geq c$, then **Halt** or else go to line 1;
10: **else**
11:　　Go to line 3;
12: **end if**

7.4 Experimental Methods

This section briefly introduces the typical experimental methods, including *benchmark datasets*, *learning objectives* and *computing frameworks*.

7.4.1 Benchmark Datasets

There are some benchmark datasets commonly used in differentially private deep learning domain. These include:

- MNIST[1] [133]. This dataset consists of 60,000 training samples and 10,000 testing samples of handwritten digit recognition. Each example is a 28 × 28 image. Both [202] and [6] evaluated their the proposed algorithms using this dataset.
- CIFAR [47]. The CIFAR dataset comprises labeled subsets of the 80 million Tiny Images[2] dataset. There are two versions of different numbers of image classes. The CIFAR-10 dataset that consists of 50,000 training samples and 10,000 test samples belonging to ten classes. The CIFAR-100 dataset also has 50,000 training samples, but they are separated into 100 classes. Abadi et al. [6] mainly used the CIFAR-10 dataset to conduct the experiments and used the CIFAR-100 dataset as a public dataset to train the network.
- SVHN[3] [170]. This dataset is a collection of the house numbers from Google Street View images. It contains over 600,000 samples; each sample is a 32 × 32 image with a single character in the center. Shokri et al. [202] used 100,000 samples for training and 10,000 samples for each test.

7.4.2 Learning Objectives

The learning objective stated in Shokri et al.'s work was to classify input data from the MNIST and SVHN datasets into ten classes[202]. The accuracy of the differentially private DSSGD protocol was compared to SGD and to a standalone scenario. In the work of Abadi et al.'s work [6], the classification accuracy of the proposed algorithm was evaluated on a training set and test set by changing various parameters, such as the privacy budget, the learning rate and so on. In Phanh et al.'s work [180], the learning objective was to execute the proposed deep private auto-encoder for a binomial classification task on health data from a social network. The prediction accuracy of the compared algorithms was evaluated by varying the dataset's cardinality and the privacy budget.

---

[1] http://yann.lecun.com/exdb/mnist/.

[2] http://groups.csail.mit.edu/vision/TinyImages/.

[3] http://ufldl.stanford.edu/housenumbers/.

7.4.3 Computing Frameworks

Several deep learning frameworks[4] have provided abundant resources for researchers (e.g., *TensorFlow*, *Torch7*, *Caffee*, and so on). Among existing research on differentially private deep learning, two frameworks have been commonly used to conducting experiments:

- Torch7 [48, 219] and Torch7 nn packages.[5] Torch7 is based on Torch, which is a MATLAB-like environment built on LuaJIT. It is composed of eight built-in packages. Of these, the Torch7 nn package provides standard modules for building neural networks. The work of [202] adopts a multi-layer perceptron (MLP) and a convolutional neural network (CNN) for its experiments and implements them in Torch7.
- TensorFlow.[6] TensorFlow was developed by the Google Brain Team and belongs to symbolic framework which is specified as a symbolic graph of vector operations. The work of [6] implemented their differentially private SGD algorithm using this framework.

A detailed comparison of deep learning frameworks was conducted by Anusua Trivedi in 2016.[7]

7.5 Summary

Although deep learning has achieved impressive success in many applications, the existing research on differentially private deep learning is mainly focused on classification tasks, such as the image classification tasks in [202] and [6], and the human behavior classification tasks in [180]. Many other deep learning tasks, such as language representation, may provide future opportunities for applying differential privacy to enhance related learning processes.

Remarkably, artificial neural networks may not be the only way to develop a deep learning model. The deep forest model, proposed by Zhou et al. [254] and based on a decision tree ensemble approach, has shown good properties for hyper-parameter tuning. It contains efficient and scalable training processes and has maintained excellent performance compared to traditional approaches in initial experiments on small-scale training data. This work may lead to potential privacy issues that will need to be addressed, and motivates a further novel research area for differentially private algorithm design.

---

[4]https://en.wikipedia.org/wiki/Comparison_of_deep_learning_software.

[5]https://github.com/torch/nn.

[6]https://www.tensorflow.org.

[7]http://blog.revolutionanalytics.com/2016/08/deep-learning-part-1.html.

Chapter 8
Differentially Private Applications: Where to Start?

8.1 Solving a Privacy Problem in an Application

A lot of differentially private applications have been proposed nowadays. The various steps that can be followed when solving a privacy preservation problem for a particular application are shown in Fig. 8.1. The dark boxes in the flowchart show the steps, and the orange boxes illustrate the possible choices. First, it is necessary to identify the scenarios: data publishing or data analysis. Data publishing aims to release answers to queries or entire datasets to public users; whereas, data analysis normally releases a private version of a model. Because private learning frameworks solve privacy preservation problems using optimization, an optimization objective normally has to be determined. The second step is identifying challenges in the application. Although differential privacy is considered to be a promising solution for privacy preservation issues, implementation in some applications still presents a number of challenges. These challenges, and their possible solutions, are introduced in the next subsection.

The third step is selecting the most suitable mechanisms. Some major mechanisms, such as transformation and iteration, have already been presented in Chap. 3. In the real world, a curator can develop more useful mechanisms based on the specific requirements of the application. Once the solution, or algorithm, for a problem has been determined, a utility and privacy analysis are required. As mentioned in Chap. 2, the most popular utility measurement is an error measurement, which evaluates the distance between the non-private output and the private output. Various applications may interpret these errors in different ways. For example, in recommender systems, the error could be a mean average error, while in location-based services, the error is normally a Euclidean distance error. The error is bounded by the accuracy parameter α, which can be estimated using tools like a union

© Springer International Publishing AG 2017

T. Zhu et al., *Differential Privacy and Applications*,

Advances in Information Security 69, DOI 10.1007/978-3-319-62004-6_8

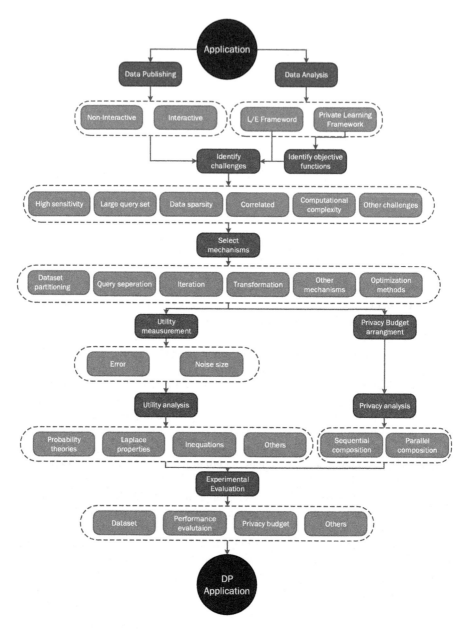

Fig. 8.1 Application flowchart

bound from probability theory, a property in a Laplace distribution, or some other useful inequality. Once the privacy budget has been assigned to each operation in the algorithm, the privacy level of the algorithm is estimated through parallel or sequential composition.

The final step is experimental evaluation, which evaluates the algorithm's performance for its intended application. Experimental evaluation involves: choosing the datasets, selecting a performance evaluation measurement, and assigning a privacy budget. Several popular public datasets are presented in the next subsection. Performance is highly associated with the utility analysis and can be considered as a practical proof of the utility analysis. Evaluating the performance of a private algorithm normally involves a comparison between the proposed algorithm, other private algorithms, and non-private algorithms. Typically, the privacy budget is set to less than 1 for each operation. The application is considered to be differentially private when all the above steps are complete.

8.2 Challenges in Differentially Private Applications

8.2.1 High Sensitivity Challenge

The most serious challenge in many applications is that queries with high sensitivity lead to high levels of noise, which may significantly decrease the utility of the applications. The noise added by differential privacy is calibrated according to the sensitivity of the query. Simple queries, such as count or sum, introduce minor noise, which has a low effect on utility. However, queries with high sensitivity are found in many real-world applications, such as the similarity measurements in recommender systems or the cluster center measurements in a clustering algorithm. In these cases, deleting one record in the dataset will have a significant impact on the similarity result. The essential challenge for differential privacy is how to decrease the noise in queries with high sensitivity.

One of the most useful ways to deal with high sensitivity is to change the definition of global sensitivity to be application-aware, which can create local sensitivity. The notion of global sensitivity, which considers worst-case scenarios, has quite a rigorous definition, but in many applications it may not be necessary to use such strict sensitivity. If using global sensitivity generates some redundant noise, the definition of sensitivity can be relaxed to achieve a trade-off between privacy and utility.

8.2.2 Dataset Sparsity Challenge

When differential privacy uses exponential or other randomization mechanisms, a sparse dataset will induce redundant noise. An exponential mechanism is an essential differential privacy mechanism that allocates different probabilities to every possible output for a query and then selects one output according to these probabilities. If the dataset is sparse, there will be many possible outputs, and a randomization mechanism will introduce a massive amount of errors compared to non-private mechanisms.

One possible solution is to shrink the domain of randomization. The noise will decrease when the range that is randomized is limited. For example, clustering is one possible way to structure existing items into groups and limit the domains to be randomized within each cluster.

8.2.3 Large Query Set Challenge

In real-world applications, such as data mining and recommendation systems, large numbers of queries have to be released. In addition, many applications have multiple steps where noise needs to be added. However, difficulties occur when the query set is large because the privacy budget needs to be divided into tiny pieces. In these scenarios, a large amount of noise will be introduced into the published query answers.

For example, suppose we have a privacy budget ϵ and the sensitivity equals 1. If algorithm A has only one step where noise is added, the total noise will be $Lap(1/\epsilon)$. But, if recursive algorithm B repeats over ten rounds and noise has to be added to each round, the privacy budget ϵ will be divided into ten pieces, and the noise added to each round will be $Lap(10/\epsilon)$. The total noise will be $10 * Lap(10/\epsilon)$, which is much higher than the noise in algorithm A. Unfortunately, most applications have more than one step where noise needs to be added. Therefore, a critical challenge in real-world applications is how to decrease noise when answering a large query set.

Several possible solutions have been proposed in previous research. One possible solution is to apply an iteration mechanism. Adding noise to each step in a recursive algorithm is accomplished using an iteration mechanism, and, given that some noise will be generated without consuming the privacy budget, the total amount of noise will be reduced. Iteration mechanisms were discussed in Chap. 3. Another possible solution is to release a synthetic dataset, which can be used to answer multiple queries without consuming any more of the privacy budget. Synthetic dataset publishing was discussed in Chap. 5. In addition, a parallel composition can be applied when the query is only performed on a subset of the dataset.

8.2.4 Correlated Data Challenge

Existing research on differential privacy assumes that, in a dataset, data records are sampled independently. However, in real-world applications, data records are rarely independent [32]. When a traditional differentially private technique performed on a correlated dataset discloses more information than expected, this indicates a serious privacy violation [126]. Recent research has been concerned with this new kind of privacy violation; however, coupled datasets lack a solid solution. Moreover, how to decrease the large amount of noise incurred by differential privacy in a correlated dataset remains a challenge.

As advances in correlated data analysis are made, it may now be possible to overcome these research barriers. Correlated information can be modeled by functions or parameters that can be further defined as background information in a differential privacy mechanism. For example, Cao et al. [32] use time intervals and correlation analyses to identify correlated records, and they model correlated information using inter-behavior functions. This solution may help tackle these research barriers by incorporating modeled information into a differential privacy mechanism as most records are only partially correlated. In other words, the impact of deleting a particular record may differ among the remaining records. If those differences can be quantified, they can be used to calibrate the noise. New sensitivity can be defined by correlating the quantified differences.

8.2.5   Computational Complexity Challenge

Differential privacy introduces extra computational complexity to applications. Complexity is not a significant issue in traditional Laplace mechanisms; however, in other mechanisms, such as iteration mechanisms or mechanisms that need to traverse the whole universe, computational complexity is exponential to the size of the universe. This is a substantial obstacle when implementing differential privacy in various applications, especially in real-time systems.

The most effective solution for eliminating the impact of computational time is to avoid traversing the entire universe. The curator can partition the domain into smaller subsets to eliminate the search space, or convert the search domain into a tree structure to decrease complexity. Thus, sampling is a useful technique for decreasing search times.

8.2.6   Summary

In summary, these challenges prevent differential privacy from being applied to a broader range of real-world applications. Current research is trying to fix these challenges in several practical ways. Table 8.1 summarizes these challenges and their possible solutions, which can be used in various applications.

Table 8.1 Challenges and possible solutions

Challenges	Possible solutions	Disadvantages
High sensitivity	Use local sensitivity	Sacrifice privacy; local sensitivity may not be easy to estimate
	Change the function	May sacrifice utility, not easy to replace a high sensitivity query
Large query set	Iterative based mechanism	The parameters are not easy to adjust; introduce extra computational complexity
	Publish synthetic dataset	Only useful for specific target; introduce extra computational complexity
	Apply privacy composition theorem	May not be applicable for some applications
	Break the correlation between queries	May not be applicable for some applications
Data sparsity	Shrink the scale of the dataset	Shrinking process may introduce extra noise; sacrifice privacy level
Correlated data	Define correlated degree and use local sensitivity	Only suitable for data with lower correlation
Computational complexity	Sampling	May loss some important information
	Avoid traverse the entire universe	May not be applicable for some applications

8.3 Useful Public Datasets in Applications

8.3.1 Recommender System Datasets

The Netflix dataset[1] is a real industrial dataset released by Netflix. It was extracted from the *Netflix* Prize dataset. Each user has rated at least 20 movies, and each movie has been rated by 20–250 users.

The MovieLens[2] datasets are the benchmark datasets for recommender system research. GroupLens Research has collated movie rating datasets of various sizes to cater for diverse requirements. For example, the MovieLens 20M dataset is the latest dataset and provides 20 million ratings and 465,000 tag applications applied to 27,000 movies by 138,000 users. It includes tag genome data with 12 million relevance scores across 1100 tags.

The Yahoo! datasets[3] represents a snapshot of the Yahoo! Music community's preferences for various songs. The dataset contains over 717 million ratings of 136,000 songs provided by 1.8 million users of Yahoo! Music services. The data were collected between 2002 and 2006.

---

[1] http://www.netflixprize.com.

[2] http://www.grouplens.org.

[3] https://webscope.sandbox.yahoo.com/catalog.php?datatype=r.

8.3.2   Online Social Network Datasets

The Stanford Network Analysis Platform (SNAP) is a general purpose network analysis and graph mining library that includes datasets describing very large networks, including social networks, communications networks, and transportation networks [136].

8.3.3   Location Based Datasets

The *T-Drive* trajectory dataset contains 1-week of trajectories for 10,357 taxis. It includes around 15 million total data points with a total trajectory distance reaching 9 million kilometers.

GeoLife is a location-based social networking service, enabling users to share their life experiences and build connections using their GPS location history. This dataset includes 178 users over a period of 4 years (from April 2007 to October 2011). It contains 17,621 trajectories covering a distance of 1,251,654 km, and a total duration of 48,203 h.

Gowalla is a location-based social networking website where users checking-in to share their location. This dataset contains a total of 6,442,890 user check-ins of the users during the period between February 2009 and October 2010. The main objective behind the use of this dataset is to understand the basic laws that govern the human motion and dynamics. This dataset belongs to SNAP.

There are also several well-known websites that can crawl photos using GPS information, for example, *Flickr*: www.flickr.com. *Instagram*: http://instagram.com.

8.3.4   Other Datasets

The *UCI Machine Learning repository*[4] is a collection of datasets, domain theories, and data generators that are used by the machine learning community to empirically analyze machine learning algorithms. It provides a number of datasets that can be used to evaluate privacy preservation methods. For example, the Adult dataset from the repository was originally used to predict whether a citizen's income would exceed 50,000 p.a. based on census data. It is also known as the "Census Income" dataset and is a popular dataset used to evaluate privacy preservation methods.

---

[4]http://archive.ics.uci.edu/ml/.

8.4   Applications Settings

The basic settings of an application based on the flowchart in Fig. 8.1 are illustrated
in Table 8.2. This nomenclature is used in subsequent chapters.

The first row shows the type of applications, for example location privacy, recommender system or social network. The nature of the input and output data is also
listed in the table. The input data, output data and utility measurements are highly
related to the application. For example, the input data of a recommender system are
normally a user-item matrix, while social network data are normally graphs. The
output data of a recommender system are normally rating predictions, while social
network output data can be statistics information or graphs. Challenges and possible
solutions vary across application types. The publishing setting can be interactive or
non-interactive, while the analysis setting can be a Laplace/exponential framework
or a private learning framework. Utility measurements are normally based on the
error measurement, which is the distance between the non-private result and the
private result. For example, in a recommender system, the utility can be measured
by the distance between the original MAE and the MAE based on the private result.

Table 8.2 Application settings

Application	Recommender systems, location privacy...
Input data	Transaction, graph, stream, ...
Output data	Prediction, statistics information, synthetic data, ...
Publishing/analysis setting	Interactive; non-interactive; Laplace/exponential framework; private learning framework...
Challenges	High sensitivity, ...
Solutions	Adjust sensitivity measurement...
Selected mechanism	Partition, iteration, query separation, ...
Utility measurement	Error measurement
Utility analysis	Probability theories, Laplace properties, inequations,...
Privacy analysis	Sequential composition, parallel composition
Experimental Evaluation	MAE, RMSE, etc.

Chapter 9
Differentially Private Social Network Data Publishing

9.1 Introduction

With the significant growth of Online Social Networks (OSNs), the increasing volumes of data collected in those OSNs have become a rich source of insight into fundamental societal phenomena, such as epidemiology, information dissemination, marketing, etc. Much of this OSN data is in the form of graphs, which represent information such as the relationships between individuals. Releasing those graph data has enormous potential social benefits. However, the graph data infer sensitive information about a particular individual [13], which has raised concern among social network participants.

If the differential privacy mechanism is adopted in graph data, the research problem is then to design efficient algorithms to publish statistics about the graph while satisfying the definition of either node differential privacy or edge differential privacy. The former protects the node of the graph and the latter protects the edge in the graph.

Previous works have successfully achieved either node differential privacy and edge differential privacy when the number of queries is limited. For example, Hay et al. [95] implemented node differential privacy, and pointed out the difficulties to achieve the node differential privacy. Paper [229] proposed to publish graph dataset using a *dK-graph* model. Chen et al. [41] considered the correlation between nodes and proposed a correlated release method for sparse graphes. However, these works suffer from a serious problem: when the number of queries is increasing, a large volume of noise will be introduced. This problem can be tackled by iteration mechanism, which will be presented in this chapter.

This chapter focuses on both node and edge differential privacy. First, the chapter present several basic methods on node and edge differential privacy, then proposed an iteration method to publish a synthetic graph. Specifically, given a set of queries, an iteration method is used to generate a synthetic graph to answer these queries accurately. The iteration process can be considered as a training procedure, in

© Springer International Publishing AG 2017
T. Zhu et al., *Differential Privacy and Applications*,
Advances in Information Security 69, DOI 10.1007/978-3-319-62004-6_9

Table 9.1 Application settings

Application	Social network data publishing
Input data	Graph
Output data	Query answers and synthetic graph data
Publishing setting	Non-interactive
Challenges	Large query set
Solutions	Iterative based method
Selected mechanism	Iteration
Utility measurement	Noise measurement
Utility analysis	Laplace properties
Privacy analysis	Sequential composition
Experimental evaluation	Graph distance

Fig. 9.1 Graph

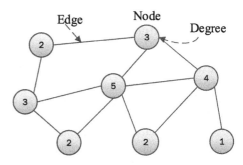

which queries are training samples and the synthetic graph is an output learning model. A synthetic graph is finally generated by iterative update to answer the set of queries. As the training process consumes less privacy budget than the state-of-the-art methods, the total noise will be diminished. Table 9.1 shows the basic setting of the application.

9.2 Preliminaries

We denote a simple undirected graph as $G\langle V, E\rangle$, where $V = \{v_1, v_2, \ldots, v_n\}$ is a set of vertices (or nodes) representing individuals in the social network and $E \subseteq \{(u, v)|u, v \in V\}$ is a set of edges representing relationships between individuals. Figure 9.1 shows an example of a graph, in which the nodes are represented by circles and edges are represented by lines. The degree of a node refers to the number of its neighbourhoods. Formally, we define degree as follows:

Neighbourhood:

$$N(v) = \{u|(u, v) \in E, u \neq v\}; \tag{9.1}$$

Fig. 9.2 Queries on graphs

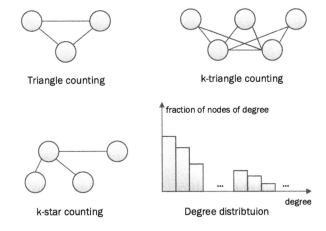

Triangle counting   k-triangle counting

k-star counting   Degree distribtuion

Degree:

$$D(v) = |N(v)|. \tag{9.2}$$

Besides the count of nodes and edges in a graph, Fig. 9.2 shows several popular subgraph queries, including triangle counting, *k-triangle* counting, *k-stars* counting, and degree distribution.

9.3   Basic Differentially Private Social Network Data Publishing Methods

9.3.1   Node Differential Privacy

Node differential privacy ensures the privacy of a query over two neighbouring graphs where two neighbouring graphs differ one node and all edges connected to the node. Figure 9.3 shows the neighboring graphes in node differential privacy. Hay et al. [95] first proposed the notion of node differential privacy and pointed out the difficulties to achieve it. They showed that the result of query was highly inaccurate for analyzing graph due to the large noise.

Let us use Fig. 9.3 to illustrate the problem. When answering query f_1: how many nodes are there in the graph? the $\triangle f_1$ equals to 1, and the noise adding to f_1 is scaled to $Lap(1/\epsilon)$, which is quite low. However, when answering query f_2: how many edges are there in the graph? the sensitivity of f_2 equals to 5 as the maximum degree of all nodes is 5. The noise adding to the f_2 result is quite large comparing with the f_1.

The high sensitivity of node differential privacy derives from the degree of a node. When deleting a node, the maximum changing is determined by the largest degree of node in a graph. Theoretically, the sensitivity of a graph G will be

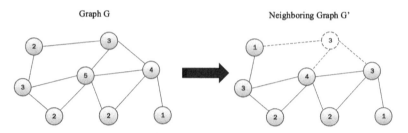

Fig. 9.3 Neighboring graph in node differential privacy

maximum to $n - 1$. How to decrease the sensitivity of f_2 is a challenge. One of the key ideas to achieve a better utility of node differential privacy is to transform the original graph to a new graph with lower sensitivity. Several methods have been proposed to achieve the goal. These methods roughly can be grouped into three categories: truncation, Lipschitz extension and iterative based mechanism.

9.3.1.1 Truncation and Smooth Sensitivity

Truncation method transforms the input graphes into graphes with maximum degree below a certain threshold θ [117]. The graph G is truncated to G_θ by discarding nodes with degree $> \theta$. Figure 9.4 shows a truncated graph G_3 for original graph G, in which one node with degree 5 is discarded to make sure all nodes are equal or below the degree of 3. By this way, the sensitivity of edge counting query will be decreased from sensitivity 5 to sensitivity 3.

Algorithm 1 ϵ-Node-Private Algorithm for Publishing Degree Distributions

---

Require: G, ϵ, θ, degree distributions query f
Ensure: $\widehat{p}$.

    1. determine the randomized truncated parameter: select $\widehat{\theta} \in \{D + \frac{\log n}{\beta} + 1, \ldots, 2D + \frac{\log n}{\beta}\}$;

    2. computer $G_{\widehat{\theta}}$ and smooth bound $S(G_\theta)$ with $\beta = \frac{\epsilon}{\sqrt{2(\widehat{\theta}+1)}}$;

    3. output $\widehat{f} = f(G_{\widehat{\theta}}) + Cauchy(\frac{2\sqrt{2}}{\epsilon}\widehat{\theta}S(G_\theta))^{\widehat{\theta}+1}$.

Kasiviswanathan et al. [117] showed that given a query f defined on the trunked graph G_θ, a smooth bound S is necessary for the number of nodes whose degrees may change due to the truncation. They applied Nissim et al.'s [172] β-smooth bound $S(G)$ for local sensitivity, which has been discussion in Definition 2.4 in Chap. 2. One can add noise proportional to smooth bounds on the local sensitivity using a variety of distributions. Kasiviswanathan et al. used the Cauchy distribution $Cauchy(\sqrt{2}S(G)/\epsilon)$. Algorithm 1 shows a typical algorithm to publish degree distributions for a G.

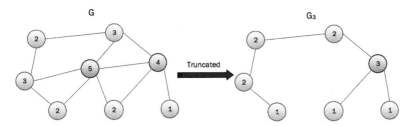

Fig. 9.4 Truncation method on graph

There are three major steps in the algorithm. The first step determines the truncated parameter θ. As we may not know the maximum degree in the graph, or the maximum degree may be very large, θ is normally approximated by $\widehat{\theta}$ Therefore, the algorithm randomized the cutoff to obtain a better bound. Given a target parameter θ, the algorithm picks a random parameter in a range of bounded constant multiple of θ. The second step creates the truncated graph by discarding the node with degree greater than θ. The final step add Cauchy distributed noise to each entry of the degree distribution.

Truncating G to G_θ is not easy, as deleting all nodes with degree greater than θ will ultimately delete more nodes and edges than we expected. Blocki et al. [27] solved this problem by selecting an arbitrary order among the edges, traversing the edges and removing each encountered edge that is connected to a node that has degree is greater than θ. Day [53] used an reverse way to create truncated graph. They first deleted all edges and add edges in a pre-defined order to achieve G_θ.

9.3.1.2   Lipschitz Extension

A more efficient method to achieve node differential privacy is based on Lipschitz extension. A function f' is a *Lipschitz extension* of f from G_θ to G if it satisfies with (1) f' agrees with f on G_θ, and (2) the global sensitivity of f' on G is equal to the local sensitivity of f on G_θ. As the sensitivity of f' is lower than that of f, Lipschitz extension of f is considered as an efficient way to decrease the sensitivity.

Figure 9.5 shows two conditions of Lipschitz extension. The large square is used to show all graphes with all possible degrees, and the eclipse inside the square is applied to show G_θ. For a query f, the global sensitivity is denoted by Δf, which is larger than the local sensitivity Δf_θ. If the algorithm can find an efficiently computable Lipschitz extension f' that is defined on all of G, then we can use the Laplace mechanism to release $\widehat{f}(G)$ with relatively small additive noise. Consequently, the target of the algorithm is to find a f' for f.

Kasiviswanathan et al. [117] proposed a flow-based method to implement such extensions for subgraph counts. Figure 9.6 shows the graph flow. Given a graph $G = (V, E)$ and a degree bound θ, the flow-based method first constructs a flow graph by copy two versions of nodes set $V_l = v_l | v \in V$ and $V_r = v_r | v \in V$, which are called left and right copies, respectively. The flow graph of G with a source s and a sink t is a directed graph on nodes $V_l \bigcup V_r \bigcup \{s, t\}$. Edges (s, v_l) and (v_r, t)

Fig. 9.5 Lipschitz extension

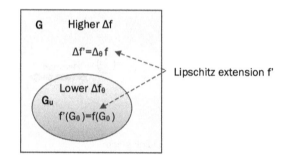

Fig. 9.6 Flow based method
to obtain Lipschitz extension

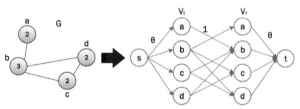

are with capacity θ, while each edge (u, v) in E is added as (u, v') between v_l and v_r with capacity 1. Let $v_{fl}(G)$ denote the value of maximum flow from s to t, $v_{fl}(G)/2$ is a Lipschitz extension of an edge query f. The global node sensitivity $\triangle v_{fl}(G) \leq 2\theta$. Accordingly, Kasiviswanathan et al. [117] published the number of edges by Algorithm 2.

Algorithm 2 ϵ-Node-Private Algorithm for Publishing Edge Numbers

---

Require: G, ϵ, θ, number of edge query f
Ensure: $\widehat{f}$.
   1. $\widehat{f} = f(G) + Lap(2n/\epsilon)$ and threshold $\tau = \frac{n\log n}{\epsilon}$;
  if $\widehat{f} \geq 3\tau$ **then**
     2. output $\widehat{f}$.
  else
     3. compute $v_{fl}(G)$ with θ;
  end if
   4. $\widehat{f} = v_{fl}(G)/2 + Lap(2\theta/\epsilon)$.

Blocki et al. [27] proceeded with a similar intuition. They showed that Lipschitz extensions exist for all real-valued functions, and give a specific projection from any graph to a particular degree-bounded graph, along with smooth upper bound on its local sensitivity.

Above works can only efficiently compute Lipschitz extensions for one-dimensional functions, in which the output is a single value. Raskhodnikova et al. [189] developed Lipschitz extensions for degree distribution queries with multidimensional vector outputs, via convex programming. Specifically, they designed Lipschitz extensions with small stretch for the sorted degree list and for the degree distribution of a graph.

9.3.1.3 Iterative Based Mechanism

Chen et al. [45] proposed an iterative based method to achieve node differential privacy. Given graph G and any real-valued function f, they defined a sequence of real-valued functions $0 = f_0(G) \leq f_1(G) \leq \cdots \leq f_m(G) = f(G)$ with the *recursive monotonicity* property that: $f_i(G') \leq f_i(G) \leq f_{i+1}(G')$ for all neighbors G and G' and $\forall i \in \{0, 1, \cdots, m\}$. They then defined quantity X to approximate the true answer of f, and the global sensitivity of X is Δ. $X_\Delta(G) = \min_{i \in [0,1]}(f_i(G) + (n - i)\Delta)$, where $X_\delta(G) \leq f(G)$ but close to $f(G)$ for larger Δ. For a carefully chosen Δ, they output the $X_\Delta(G)$ via Laplace mechanism in the global sensitivity framework, as an approximation of the real-valued function $f(G)$. This recursive approach can potentially return more accurate subgraph counting for any kinds of subgraphs with node differential privacy. However, constructing the sequence of functions $f_i(G)$ is usually NP-hard, and how to efficiently implement it remains an open problem.

9.3.2 Edge Differential Privacy

Edge differential privacy means adding or deleting a single edge between two nodes in the graph makes negligible difference to the result of the query. The first research over edge differential privacy was conducted by Nissim et al. [172], who showed how to evaluate the number of triangles in a social network with edge differential privacy. They showed how to efficiently calibrate noise for subgraph counts in terms of the smooth sensitivity. The results of this technique are investigated by Karwa et al. [113] to release counts on k-triangles and k-stars.

Rastogi et al. [190] studied the release of more general subgraph counts under a much weaker version of differential privacy, edge adversarial privacy, which considers a Bayesian adversary whose prior knowledge is drawn from a specified family of distributions. By assuming that the presence of an edge does not make the presence of other edges, they computed a high probability upper bound on the local sensitivity, and then added noise proportional to that bound. Rastogi et al.'s method can release more general graph statistics, but their privacy guarantee protects only against a specific class of adversaries, and the magnitude of noise grows exponentially with the number of edges in the subgraph.

Hay et al. [95] considered releasing a different statistics about graph, the degree distributions. They showed that the global sensitivity approach can still be useful when combined with post-processing of the released output to remove some added noise, and constructed an algorithm for releasing the degree distribution of a graph, with the edge differential privacy.

Based on the local sensitivity, Zhang [249] adopted exponential mechanism to sample the most suitable answer for subgraph query. The score function is designed carefully to make sure it reflects the distribution of query outputs. A different approach was proposed by Xiao [234], who inferred the networks structure via connection probabilities. They encoded the structure information of the social network by the connection probabilities between nodes instead of the presence or absence of the edges, which reduced the impact of a single edge.

Zhu et al. [260] proposed an iterative graph published method to achieve edge differential privacy. They proposed a graph update method that transfers the query publishing problem to an iteration learning process. The details are presented in the following section.

9.4 Graph Update Method

9.4.1 Overview of Graph Update

The proposed method is called *Graph Update* method as the key idea is to update a synthetic graph until all queries have been answered [260]. For a social network graph G and a set of queries $F = \{f_1, \ldots, f_m\}$, the publishing goal is to release a set of query results $\widehat{F}$ and a synthetic graph $\widehat{G}$ to the public. The general idea is to define an initial graph $\widehat{G}_0$ and update it to $\widehat{G}_{m-1}$ in m round according to m queries in F. Release answers $\widehat{F}$ and the synthetic graph $\widehat{G}$ are generated during the iteration. During the process, four different types of query answer involve in the iteration:

- *True answer a_t*: this is the real answer that a graph response to a query. True answer can not be published directly as it will arise privacy concern. The true answer is normally used as the baseline to measure the utility loss of a privacy-preserving algorithm. The symbol a_t is used to represent the true answer for a single query f, and $A_t = F(G) = \{a_{t1}, \ldots, a_{tm}\}$ is applied to represent an answer set for a query set F.
- *Noise answer a_n*: when we add Laplace noise to a true answer, the result will be the noise answer. Traditional Laplace method will release the noise answer directly. However, as we mentioned in Sect. 9.1, it will introduce large amount of noise to the release result. A single query answer is represented by $a_n = \widehat{f}(G) = f(G) + Lap(s/\epsilon)$ and an answer set is represented by $A_n = \widehat{F}(G) = \{a_{n1}, \ldots, a_{nm}\}$.
- *Synthetic answer a_s*: this is the answer generated by a synthetic graph $\widehat{G}$. A single query is presented by $a_s = f(\widehat{G})$ and $A_s = F(\widehat{G}) = \{a_{s1}, \ldots, a_{sm}\}$ is applied to represent an answer set.
- *Release answer a_r*: this is the answer finally released after the iteration. In Graph Update method, the release answer set will consist of noise answers and synthetic answers. The algorithm applied $a_r = \widehat{f}$ and $A_r = \widehat{F} = \{a_{r1}, \ldots, a_{rm}\}$ to represent the single answer of a query and the answer set, respectively.

These four different query answers will control the graph update process. The overview of the method is presented in Fig. 9.7. On the left side of the figure, the

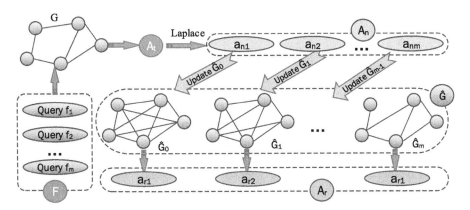

Fig. 9.7 Overview of *Graph Update* method

query set F performs on the G to get true answer set A_t. Laplace noise is then added to A_t to get a set of noise answer $A_s = \{a_{s1}, \ldots a_{sm}\}$. Each noise answer a_{si} helps to update the initial $\widehat{G}_0$ and produce a release answer a_{ri}. The method eventually outputs $A_r = \{a_{r1}, \ldots, a_{rm}\}$ and the $\widehat{G}_m$ as final results.

Comparing with the traditional Laplace method, the proposed *Graph Update* method adds less noise. As some queries are answered by the synthetic graph, these query answers will not consume any privacy budget. Moreover, the synthetic graph can be applied to predict new queries without any privacy budget. Eventually, the *Graph Update* method can outperform the tractional Laplace method.

Algorithm 3 Graph Update Method
___

Require: $G, F = \{f_1, \ldots, f_m\}, \epsilon, \eta_0$
Ensure: $A_r = \{a_{r1}, \ldots, a_{rm}\}$.
    1. $\epsilon' = \epsilon/m$
    2. initial graph $\widehat{G}_0$;
  for each query $f_i \in F$ **do**
      3. Compute true answer a_{ti};
      4. Add Laplace Noise to true answer $a_{ni} = \widehat{f}_i = f_i(G) + Lap(S/\epsilon')$;
      5. Compute synthetic answer $a_{si} = f_i(\widehat{G})$;
      6. $\eta_i = a_{ni} - a_{si}$;
    if $|ta_i| > \eta_0$ **then**
        7. $a_{ri} = a_{ni}$;
        8. update the $\widehat{G}_{i-1}$ to $\widehat{G}_i$;
    else
        9. $a_{ri} = a_{si}$;
        10. $\widehat{G}_m = \widehat{G}_{m-1}$
    end if
  end for
    11. Make all degrees in G round numbers.
    12. Output $A_r = \{a_{r1}, \ldots, a_{rm}\}$, and $\widehat{G}$;
___

9.4.2 Graph Update Method

At a high level, the *Graph Update* method works in three steps:

- *initial the synthetic graph*: As the method only preserves the edge privacy, it assumes that the number and the labels of nodes are fixed. The synthetic graph is initialed as a fully connected graph with fixed nodes.
- *update the synthetic graph*: the initial graph will be updated according to result of each query in F, until all queries in F have been used.
- *release query answers and synthetic graph*: Two types of answers, noise answers and synthetic answers that have potential to be released. Synthetic graph is also released to the public.

Algorithm 3 is a detailed description of the *Graph Update* method. In step 1, the privacy budget ϵ is divided by m and will be arranged to each query in the set. Step 2 initializes the graph to $\widehat{G}_0$ as a full connected one. Then for each query f_i in the query set F, the algorithm computes the true answer $f_i(G)$ at Step 3. After that, the noise answer and the synthetic answer of f_i are computed at Step 4 and 5, respectively. Step 6 measures the distance between the true answer and the synthetic answer. If the distance is larger than a threshold η_0, the Step 7 will release the noisy answer. Otherwise, the synthetic graph will be updated by an *Updated Function* in Step 8 and Step 9 will release the synthetic answer. This means the synthetic graph is applicable for answering question, so in Step 10, the algorithm puts the current synthetic graph to the next round. This process is iterated until all queries in F are preceded. Finally, As the number of edges should be a integer, the algorithm round the number of degrees in Step 11. the algorithm generates A_r and $\widehat{G}$ as the output in Step 12.

Algorithm 4 Update Function

Require: $\widehat{G}, f, \eta, \theta, (0 < \theta < 1)$
Ensure: $\widehat{G'}$.
    1. Identify related nodes V_f that f involved;
  if $\eta > 0$ **then**
      2. $D(V_f) = (1 + \theta) * D(V_f)$;
  else
      3. $D(V_f) = \theta * D(V_f)$;
  end if
    4. $\widehat{G'} = G \cup D(V_f)$.
    5. Output $\widehat{G'}$.

The parameter η_0 is a threshold controlling the distance between A_n and A_s. A larger η_0 means less update of the graph and most of the answer in A_r are synthetic answers. It leads to less privacy budget consuming, however, when the synthetic graph is far away from the original graph, the performance may not optimal. A smaller η_0 means the algorithm has more updates of the graph and most of the answer in A_r are noise answers. More privacy budgets will be consumes in this configuration. Consequently, the choice of η_0 will have impact on different scenarios.

9.4.3 Update Function

Step 8 in Algorithm 3 involves with an *Update Function*, which updates the synthetic graph $\widehat{G}$ to graph $\widehat{G}'$ according to query answers. Specifically, *Update Function* is controlled by the distance η between the a_n and a_s of f. If a_n is smaller than a_s, it means that the synthetic graph has more edges than the original graph in the related nodes. *Update Function* has to delete some edges between the related nodes. Otherwise, *Update Function* will add some edges in the synthetic graph.

These related nodes is defined in the follow Definition 9.1:

Definition 9.1 (Related Node) For a query f and a graph G, *related nodes V_f* are all nodes that response to the query f, $D(V_f)$ is used to denote degrees of those nodes.

The number of edges for a node should be a integer. However, to adjust degree of those related nodes, we arrange weight θ ($0 \le \theta \le 1$) for each edge. After the updating, these weights will be rounded to represent node edges.

Algorithm 4 illustrates the detail of *Update Function*. In the first step, the function identifies related nodes. If $\eta > 0$, which means the synthetic graph has less edges than the original one, the function will enhance the θ in Step 2. If $\eta \le 0$, which means the synthetic graph has too many edges, the function will diminish those edges by θ in Step 3. Step 4 merges the edges to the original graph. Step 5 outputs the $\widehat{G}'$.

9.4.4 Privacy and Utility Analysis

9.4.4.1 Privacy Analysis

To analyze the privacy level of the proposed method, the sequential composition is applied. For the traditional Laplace method, when answering F with m queries, ϵ will be divided into m pieces and arranged to each query $f_i \in F$. Specifically, we have $\epsilon' = \epsilon/m$ and for each query, the noise answer will be $a_{ni} = f_i + Lap(s/\epsilon')$. According to the *sequential composition*, the Laplace method preserve $(\epsilon' * m)$-differential privacy, which is equal to ϵ-differential privacy.

In *Graph Update* method, the release answer set A_r are the combination of noise answers A_n and synthetic answers A_s. Only A_n consume privacy budget, while A_s do not. In Algorithm 4, even Step 4 adds Laplace noise to the true answer, the noise result does not release directly. Only when the algorithm processed to Step 7, in which a_n is released, the algorithm consumes the privacy budget. Suppose there are $j(0 \leq j \leq m)$ queries in F is released by synthetic answers, the algorithm preserves $((m-j) * \epsilon')$-differential privacy. As $(m-j) * \epsilon' \leq m * \epsilon'$, the *Graph Update* method preserve more strict privacy than tractional Laplace method.

9.4.4.2 Utility Analysis

Error measurement is applied to evaluate the utility. The error is defined by *Mean Absolute Error* (MAE). MAE_r of release answer A_r is defined as Eq. (9.3)

$$MAE_r = \frac{1}{m}|\widehat{F}_i(G) - F_i(G)|$$

$$= \frac{1}{m}\sum_{f_i \in F}|\widehat{f}_i(G) - f_i(G)|$$

$$= \frac{1}{m}\sum_{a_i \in A_r}|a_{ri} - a_{ti}|$$

$$= \frac{1}{m}|A_r - A_t|. \tag{9.3}$$

Similarly, MAE_n of noise answers and MAE_s of synthetic answers are defined as Eqs. (9.4) and (9.5), respectively.

$$MAE_n = \frac{1}{m}|A_n - A_t|; \tag{9.4}$$

$$MAE_s = \frac{1}{m}|A_s - A_t|. \tag{9.5}$$

It is obvious that for true answers A_t, the *MAE* is zero. MAE_n represents the performance of traditional Laplace method. A lower *MAE* implies a better performance.

The target of *Graph Update* method is to achieve a lower MAE_r in a fixed privacy budget. A simulated figure, Fig. 9.8, is applied to illustrate the relationship between *MAE* values and the size of the query set m.

In Fig. 9.8, x axis is the size of the query set and y axis is the value of *MAE*. For noise answer A_n, MAE_n is arising with the increasing of m. A smooth line is applied to represent the MAE_n in this simulated figure. In real case, the line is fluctuated as the noise is derived from Laplace distribution. The MAE_s is decreasing at the beginning with the increasing of m. When it reaches to its lowest point, the MAE_s begins to rise with the enhance of m. This is because with the update of the graph, the synthetic graph is getting more and more accurate, MAE_s is keeping decreasing. However, as the iteration procedure is controlled by the noise answer, it is impossible for synthetic graph to equal to the original graph, no matter how large m is. On the contrary, with the increasing of m, more noise will be introduced to iteration and the synthetic graph will be far away from the original graph.

Fig. 9.8 Utility of the query
set on a graph

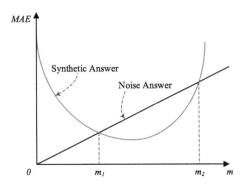

As A_r is the combination of A_n and A_s, MAE_r of release answers can be reflected by synthetic answer MAE_s and noise answer MAE_n. Figure 9.8 shows that MAE_s will below MAE_n when the query size reaches to m_1. After reaching to a lowest point, it begins to increase. After reaching to m_2, the MAE_s is higher than MAE_n. Consequently, when m in the scale of $[0, m_1) \cup (m_2, m]$, the MAE_r is dominated by noise answer MAE_n. When m in the scale of $[m_1, m_2]$, the MAE_r is dominated by synthetic answer MAE_s. By this way, in the scale of $[0, m]$, the MAE_r of release answers is smaller than MAE_n, which means that the performance of the proposed *Graph Update* method is better than the traditional Laplace method.

9.4.5 Experimental Evaluation

This section evaluates the performance of the proposed Graph Update method comparing with Laplace mechanism.

9.4.5.1 Datasets and Configuration

The experiment involve with four datasets listed in Table 9.2. These datasets are collected from Stanford Network Analysis Platform (SNAP) [136].

The experiment considers the degree query on nodes, which is similar to the count query on relation dataset. To preserve the edge privacy, the degree query has the sensitivity of 1, which means deleting an edge will have maximum impact of 1 on the query result. The performance of results is measured by *Mean Absolute Error* (MAE) (9.3).

Table 9.2 Graph datasets

Type	Name	Nodes	Edges
Social networks	Ego-Facebook	4039	88,234
Social networks	Wiki-Vote	7115	103,689
Internet peer-to-peer networks	p2p-Gnutella08	6301	20,777
Collaboration networks	ca-GrQc	5242	14,496

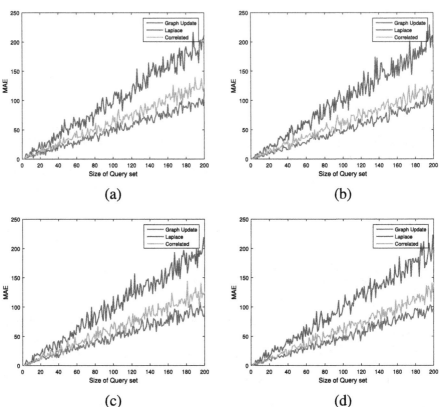

Fig. 9.9 Performance of different methods. (**a**) `ego-Facebook`. (**b**) `Wiki-Vote`. (**c**) `p2p-Gnutella08`. (**d**) `ca-GrQc`

9.4.5.2 Performance Evaluation on Diverse Size of Query Sets

The performance of the *Update Graph* is examined through comparison with the Laplace method [62] and Correlated method [41]. The size of query sets is from 1 to 200, in which each query is independent to each other. Parameters η_0 and θ as optimal ones for each dataset and the ϵ is fixed at 1 for all methods.

According to Fig. 9.9, It is observed that with the increasing of the size of the query sets, *MAE*s of all methods are increasing approximately in linear. This is because the queries are independent to each other and the privacy budget is arranged equally to each query. With the linear increasing of the query number, the noise added to each query answer is enhanced linearly.

Second, Fig. 9.9 shows that *Update Graph* has lower *MAE* comparing with other two methods, especially when the size of the query set is large. As shown in Fig. 9.9a, when the size of query set is 200, the *MAE* of *Graph Update* is 99.8500 while the Laplace method has *MAE* of 210.0020, and the *Correlated* method has *MAE* of 135.2078 which is 52.45 and 26.15% higher than the proposed *Update Graph*. This trend can be observed in Fig. 9.9b–d. The proposed *Graph Update* mechanism has better performance because part of query answers does not consume any privacy budget, while noise is only added in the updated procedure. Other methods, including Laplace method consume the privacy budget when answering every query. The experimental results show the effectiveness of *Graph Update* in answering a large set of queries.

Third, it is worth to mention that when the size of the query set is limited, the proposed *Graph Update* may not necessary outperform the *Correlated* method. Figure 9.9a shows that when the size is less than 20, *MAE*s of *Graph Update* and the *Correlated* method are mixed together. This is because when the query set is limited, the synthetic graph can not be fully updated and may differ from the original graph largely. Therefore, the performance may not necessary outperform other methods significantly. This result shows that *Graph Update* is more suitable in scenarios that need to answer a large amount of queries.

9.5 Summary

Nowadays, the privacy problem have aroused peoples attention. Especially the online social network data, which contains a massive personal information. How to release social network data is a hot topic that attracts lots of attention. This chapter proposes several method to meet with the graph publishing problem. And to overcome the problem of providing accurate results even when releasing large numbers of queries, this chapter then presents an iterative method that transfers the query release problem into an iteration based update process, so as to providing a practical solution for publishing a sequence of queries with high accuracy. In the future, much more complied queries should be investigated, such as cut queries and triangle queries, which can allow researchers to get more information of the dataset while still can guarantee users' privacy.

Chapter 10
Differentially Private Recommender System

10.1 Introduction

A recommender system attempts to predict a user's potential likes and interests by analyzing the user's historical transaction data. Currently recommender systems are highly successful on e-commerce web sites capable of recommending products users will probably like. *Collaborative Filtering* (CF) is one of the most popular recommendation techniques as it is insensitive to product details. This is achieved by analyzing the user's historical transaction data with various data mining or machine learning techniques, e.g. k nearest neighbor rule, the probability theory and matrix factorization [194]. However, there is potential for a breach of privacy in the recommendation process.

The literature has shown that continual observation of recommendations with some background information makes it possible to infer the individual's rating or even transaction history, especially for the *neighborhood-based* methods [30]. For example, an adversary can infer the rating history of an active user by creating fake neighbors based on background information [30].

Typically, a collaborative filtering method employs certain traditional privacy preserving approaches, such as cryptographic, obfuscation and perturbation. Among them, *Cryptographic* is suitable for multiple parties but induces extra computational cost [31, 247]. *Obfuscation* is easy to understand and implement, however the utility will decrease significantly [23, 176]. *Perturbation* preserves high privacy levels by adding noise to the original dataset, but the magnitude of noise is subjective and hard to control [181]. Moreover, these traditional approaches suffer from a common weakness: the privacy notion is weak and hard to prove theoretically, thus impairing the credibility of the final result. In order to address these problems, differential privacy has been proposed.

© Springer International Publishing AG 2017   107
T. Zhu et al., *Differential Privacy and Applications*,
Advances in Information Security 69, DOI 10.1007/978-3-319-62004-6_10

Differential privacy was introduced into CF by McSherry et al. [156], who pioneered a study that constructed the private covariance matrix to randomize each user's rating before submitting to the system. Machanavajjhala et al. [152] presented a graph link-based recommendation algorithm and formalized the trade-off between accuracy and privacy. Both of them employed Laplace noise to mask accurate ratings so the actual opinions of an individual were protected.

Although differentially privacy is promising for privacy preserving CF due to its strong privacy guarantee, it still introduce large noise. Large noise occurs for two reasons: the high sensitivity and the naive mechanism. Sensitivity determines the size of the noise to be added to each query. Unfortunately, the queries employed in recommendation techniques always have high sensitivity, followed by the addition of large noise. A naive mechanism is another issue that leads to high noise. Previous work directly uses the differential privacy mechanism and disregards the unique characteristics of recommendations, thus negatively affecting the recommendation performance.

To overcome the weakness, two research issues need to be considered.

- *How to define sensitivity for recommendation purposes?* Traditional global sensitivity measurement is not suitable for CF due to high dimensional input. How to define a new sensitivity is an issue to be addressed.
- *How to design the recommender mechanism for CF?* For example, the performance of neighborhood-based methods is largely dependent on the quality of selected neighbors. How to enhance the quality of selected neighbors in a privacy preserving process is another research issue. By re-designing the private selection mechanism, One can retain the accuracy from the final output result.

To achieve the goal, the chapter first proposes two typical differentially private recommender systems, and then present a differentially private neighborhood-based recommender system to defend a specific attack, KNN attack, in detail. Table 10.1 illustrates the setting of differentially private recommender system.

Table 10.1 Application settings

Application	Recommender systems
Input data	User-item rating matrix
Output data	Prediction
Publishing setting	Interactive
Challenges	High sensitivity
Solutions	Adjust sensitivity measurement
Selected mechanism	Group large candidate sets
Utility measurement	Error measurement
Utility analysis	Union bound
Privacy analysis	Parallel composition
Experimental evaluation	Performance measured by MAE

10.2 Preliminaries

10.2.1 Collaborative Filtering

Collaborative Filtering (CF) is a well-known recommendation technique that can be further categorized into *neighborhood-based methods* and *model-based methods* [193]. The neighborhood-based methods are generally based on the k nearest neighbor rule (KNN), and provides recommendations by aggregating the opinions of a user's k nearest neighbors, while model-based methods are developed using different algorithms, such as singular value decomposition, probabilistic latent semantic, etc., to predict users' rating of unrated items [210].

Let $U = \{u_1, u_2 \ldots u_n\}$ be a set of users and $I = \{t_1, t_2 \ldots t_m\}$ be a set of items. The *user* × *item* rating dataset D is represented as a $n \times m$ matrix, in which r_{ui} denotes the rating that user u gave to item t_i. For each t_i, $s(i, j)$ represents its similarity with item t_j. $s(u, v)$ denotes the similarity between user u_u with u_v. $N_k(t_i)$ denotes the set of item t_i's k neighbors, and $U_{ij} = \{u_x \in U | r_{xi} \neq \varnothing, r_{xj} \neq \varnothing\}$ denotes the set of users, co-rating on both item t_i and t_j. In addition, we use (r) to denote average rating, and $\hat{r}$ to represent inaccurate rating, including predicted rating or noisy rating.

Table 10.2 is a typical user-item matrix. Users rate different movies by various scores. The target of recommender system is to predict those empty rates by neighborhood-based or model-based methods.

10.2.2 Neighborhood-Based Methods: k Nearest Neighbors

Two stages are involved in *neighborhood-based methods*: the *Neighbor Selection* and the *Rating Prediction*. Figure 10.1 shows a k nearest neighbor algorithm. In the *Neighbor Selection* stage, the similarity between any two users or any two items is estimated, and corresponds to the *user-based* methods and the *item-based* methods. Various measurement metrics have been proposed to compute the similarity. Two of the most popular ones are the *Pearson Correlation Coefficient* (PCC) and *Cosine-based Similarity* (COS) [7]. Neighbors are then selected according to the similarity.

Table 10.2 User-item matrix for a movie recommender system

User	Alien	Logan	Beauty and the beasts	The artist
Alice	3	2	4	5
Bob		3	5	4
Cathy	2	3	4	0
...	...	...		
Eva	3		4	

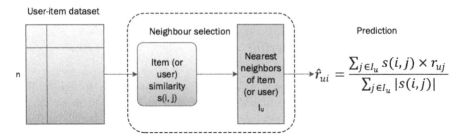

Fig. 10.1 k nearest neighbors method

- Pearson Correlation Coefficient

$$sim(i,j) = \frac{\sum_{x \in U_{ij}} (r_{xi} - \bar{r}_i)(r_{xj} - \bar{r}_j)}{\sqrt{\sum_{x \in U_{ij}} (r_{xi} - \bar{r}_i)^2} \sqrt{\sum_{x \in U_{ij}} (r_{xj} - \bar{r}_j)^2}}, \tag{10.1}$$

where $\bar{r}$ is the average rating given by relative users.
- Cosine-based Simlarity

$$sim(i,j) = \frac{r_i \cdot r_j}{||r_i||_2 ||r_j||_2}. \tag{10.2}$$

In the *Rating Prediction* Stage, for any item t_i, all ratings on t_i by users in $N_k(u_a)$ will be aggregated into the predicted rating $\widehat{r}_{ai}$. Specifically, the prediction of $\widehat{r}_{ai}$ is calculated as a weighted sum of the neighbors' ratings on item t_i. Accordingly, the determination of a suitable set of weights becomes an essential problem because most work relies on the similarity between users or items to determine the weight. For example, for *item-based* methods, the prediction of $\widehat{r}_{ai}$ is formulated as follows:

$$\widehat{r}_{ai} = \frac{\sum_{j \in I_a} s(i,j) \cdot r_{a,j}}{\sum_{j \in I_a} |s(i,j)|}. \tag{10.3}$$

In user-based methods, the active user's prediction is made by the rating data from many other users whose rating is similar to the active user. The predicted rating of user v on item α is:

$$\widehat{r}_{v\alpha} = \bar{r}_v + \frac{\sum_{u \in U_v} s(v,u)(r_{u\alpha} - \bar{r}_u)}{\sum_u |s(v,u)|},$$

where $\bar{r}_v$ and $\bar{r}_u$ is average rating given by user u and v respectively.

Finally, the computed prediction are converted into recommendations, e.g., a subset of items with the highest predicted rating is recommended to the user.

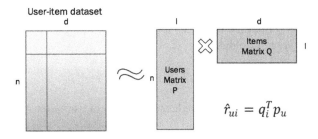

Fig. 10.2 Matrix factorization recommender system

10.2.3 Model-Based Methods: Matrix Factorization

Some of the most successful model-based methods are based on matrix factorization. Matrix factorization characterizes both items and users by vectors of factors inferred from item rating patterns. Matrix factorization models map both users and items to a joint latent factor space of dimensionality l, such that user-item interactions are modeled as inner products in that space.

Specifically, the method factorizes D into two latent matrices: the user-factor matrix P with the size of $n \times l$ and the item-factor matrix Q with the size $d \times l$. Each row p_u in P (and q_i in Q) represents the relation between the user u (item t) and the latent factor. The dimension l of the latent matrices is less than d. The dataset D is approximated as a product of P and Q, and each known rating r_{ui} is approximated by $\hat{r}_{ui} = q^T p_u$. Figure 10.2 shows the matrix factorization method.

To obtain P and Q, the method minimizes the regularized squared error for all the available ratings:

$$\min P, Q \sum_{r_{ui} \in D} [(r_{ui} - p_u \cdot q_i^T)^2 + \lambda(\|p_u\|^2 + \|q_i\|^2)]. \tag{10.4}$$

The constant λ regularizes the learned factors and prevents overfitting. Two common ways to solve the non-convex optimization problem are stochastic gradient descent (SGD) and alternating least squares (ALS).

In SGD, the factors are learned by iteratively evaluating the error $e_{ui} = r_{ui} - p_u \cdot q_i^T$ for each rating r_{ui}, and simultaneously updating the user and item vectors by taking a step in the direction opposite to the gradient of the regularized loss function:

$$p_u \leftarrow p_u + \gamma(e_{ui}q_i - \lambda p_u) \tag{10.5}$$

$$q_i \leftarrow q_i + \gamma(e_{ui}p_u - \lambda q_i). \tag{10.6}$$

The constant γ determines the rate of minimizing the error and is often referred to as the learning rate.

ALS solves the problem by updating the user and item latent vectors iteratively. In each iteration, fix P, then solve Q by least square optimization. And then fix Q, solve P by least square optimization.

Both in SGD and ALS, once the factorization converges, the latent matrices P and Q are used to predict unknown user ratings. The resulting latent vectors p_u and q_i are multiplied:

$$\widehat{r}_{ui} = q_i^T p_u, \tag{10.7}$$

which produces the predicate rating $\widehat{r}_{ui}$ of user u for item r_i.

10.3   Basic Differentially Private Recommender Systems

Differential privacy technique has been incorporated in research on recommender systems in two different settings. The first setting assumes that the recommender system is untrustworthy, so the noise will be added to the dataset before submitting to the recommender system. As the noisy data will not release any sensitivity information, the non-private recommender algorithm, including model-based and neighbourhood-based methods, can be used directly. Another setting focuses on the trusted recommender system, which can access to the user-item dataset directly. When a user applies a recommender result, the private recommender system will submit queries to user-item dataset one by one and add noise to the results. Recommendation will be performed based on the noisy result. Figure 10.3 shows two different type of settings. We present two methods to illustrate two settings in following sections.

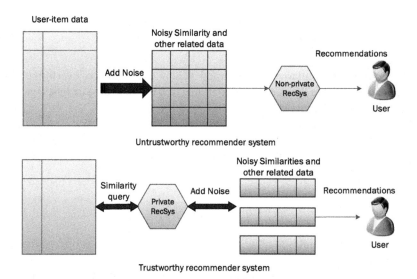

Fig. 10.3 Two settings of differentially private recommender system

10.3.1   *Differentially Private Untrustworthy Recommender System*

The first typical method is provided by McSherry et al. [156], who was the first team to introduce the differential privacy notion to collaborative filtering. They calibrated Laplace noise for each step to create a covariance matrix, and used the non-private recommender algorithm on this matrix to predict ratings. In general, it is a synthetic covariance matrix publishing method.

For a large class of prediction algorithms it suffices to have following data: $\bar{G}$: average rating for all items by all users; $\bar{I}$: average rating for each item by all users; $\bar{u}_v$: average item rating for each user; and covariance matrix (Cov). All ratings are scaled in the interval $[-B, B]$. McSherry et al.'s method consists of three steps and noise will be added to each step.

Step 1: Evaluation of Global and Item Averages, $\bar{G}$ and $\bar{I}$ Suppose there are n users with d items, the global average of item will be estimated as follows with privacy budget ϵ_1. The sensitivity of r will be the $r_{max} - r_{min} = 2B$.

$$\bar{G} = \frac{\left(\sum_D r_{ui}\right) + Lap(2B/\epsilon_1)}{|D|}. \tag{10.8}$$

Then for each item t_j, calculating the average rating for each item by all users. The average item ratings is calculated by adding a number of fictitious ratings β to stabilize the items averages, helping to limit the effect of noise for items with few ratings, while only slightly affecting the average for items with many ratings.

$$\bar{I} = \frac{\left(\sum_{D_j} r_{ui}\right) + \beta \cdot GAvg + Lap(2B/\epsilon_2)}{|D_j| + \beta}. \tag{10.9}$$

The added noise may causes the item average to go out of the range of input ratings $[r_{min}, r_{max}]$, the item average is clamped to fit this range.

Step 2: Evaluation of User Averages and Clamping of the Resulting Ratings
They follow the same technique to compute the user average ratings. The basis for evaluating the user averages is the ratings after the item averages were discounted. They stabilize the user effects with the addition of β_u fictitious ratings with the newly computed global average. The user average rating is calculated as follows. Let $D' = \{r_{ui} - \bar{I}(i) | r_{ui} \in D\}$, the adjusted global average will be

$$\bar{G}' = \frac{\left(\sum_D r'_{ui}\right) + Lap(2B/\epsilon_1)}{|D'|}.$$

Then for each user v, user average rating is

$$\bar{u}_v = \frac{(\sum_{D_v} r'_{ui}) + \beta_u \cdot \bar{G} + Lap(2B/\epsilon_2)}{|R_v| + \beta_u}.$$

The user averages are then clamped to a bounded range.

$$r_{ui} = \begin{cases} -B & \text{if } r_{ui} < -B \\ r_{ui} & \text{if } -B < r_{ui} \leq B \\ B & \text{if } r_{ui} > B. \end{cases}$$

Step 3: Calculate the Covariance Matrix The final measurement is the covariance of the centered and clamped user ratings vectors. They use per-user weights ω_u equal to the reciprocal of $\|e\|$ (the binary elements and vectors indicating the presence of ratings) as follows:

$$Cov = \sum_u w_u r_u r_u^T + b, \tag{10.10}$$

$$W = \sum_u \omega_u e_u e_u^T + b. \tag{10.11}$$

Noise b added to Cov could be large if a user's rating has large spread or if a user has rated many items. McSherry et al. provided two solutions: (1) Center and clamp all ratings around averages. If clamped ratings can be used, then the sensitivity of the function can be reduced. (2) Carefully weight the contribution of each user to reduce the sensitivity of the function. Users who have rated more items are assigned lower weights.

After the independent Gaussian noise proportional to the sensitivity bound is added to each entry in covariance matrix, all related data have been send to a non-private recommender system.

10.3.2 Differentially Private Trustworthy Recommender System

Another set of researches focus on the trusted recommender system. Friedman et al. [82] applied this setting in the matrix factorization algorithm. They claimed four different possible perturbation in the matrix factorization process, which presented in Fig. 10.4.

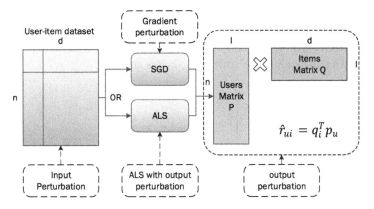

Fig. 10.4 Different perturbation methods in matrix factorization

10.3.2.1 Matrix Factorization with Private Input Perturbation

In the input perturbation approach, noise is added to each rating to maintains differential privacy. As the ratings are in the range $r_{ui} \in [r_{min}, r_{max}]$, the global sensitivity of the ratings is $\Delta r = r_{max} - r_{min} = 2B$. The input perturbation will be $\widehat{D} = r_{ui} + Lap(2B/\epsilon) | r_{ui} \in D$.

Obviously, this directly perturbation will demolish the utility of the user-item dataset. So the noisy ratings should be clamped to limit the influence of excessive noise. To mitigate this effect, they applied an additional clamping step after the introduction of noise, using the clamping parameter α, as shown in the below equation

$$
r_{ui} = \begin{cases}
-B & \text{if } r_{ui} < -B \\
0 & \text{if } |r_{ui}| \leq \alpha \\
r_{ui} & \text{if } \alpha < |r_{ui}| \leq B \\
B & \text{if } r_{ui} > B.
\end{cases}
$$

This clamping adjusts ratings in the range $[-\alpha, \alpha]$, improving the prediction accuracy.

10.3.2.2 Private Stochastic Gradient Perturbation

To perturb the SGD, the training samples are used to evaluate the prediction error resulting from the current factor matrices, and then the matrices are modified in a direction opposite to the gradient, with magnitude proportional to the learning rate γ. Suppose there are w iterations. In each iteration, for each $r_{ui} \in D$, add Laplace noise to the error e_{ui} in each iteration.

$$\widehat{e}_{ui} = r_{ui} - p_u q_i^T + Lap(2wB/\epsilon). \tag{10.12}$$

Because the error will be exceed to maximum or minimum error, clamping method will be used again to adjust the scale in the below equation

$$\widehat{e}_{ui} = \begin{cases} -e_{max} & \text{if } r_{ui} < -e_{max} \\ \widehat{e}_{ui} & \text{if } |\widehat{e}_{ui}| \le e_{max} \\ e_{max} & \text{if } e_{ui} > e_{max}. \end{cases}$$

After that, the error $\widehat{e}_{ui}$ will be used in the SGD iteration as shown in Eq. (10.5).

10.3.2.3 ALS with Output Perturbation

ALS perturbation alternately fixes one of the latent matrices P or Q, and optimize the regularized loss function for the other non-fixed matrix. When item-factor matrix Q is fixed, the overall regularized loss function can be minimized by considering for each user u the following loss function defined over the subset of ratings $D_u = \{r_{vi} \in D | v = u\}$. First, estimate the bound of p_{max} and q_{max} as $\|p_u\|_2$ and $\|q_i\|_2$, respectively. In each iteration, and for each user u in P, generate a noise vector b with $f(b) \propto exp(-\frac{\epsilon \cdot \|b\|_2}{2k} \cdot \frac{n_u \lambda}{p_{max} \cdot 2B})$. The noise vector b will be added to loss function as shown in Eq. (10.13)

$$p_u \leftarrow \arg\min_{p_u} J_Q(p_u, D_u) + b. \tag{10.13}$$

If p_u exceed the maximum value, then normalized it by p_u by Eq. (10.14)

$$p_u \leftarrow p_u \cdot \frac{p_{max}}{\|p_u\|_2}. \tag{10.14}$$

Similarly, for each item t_i in Q, sampling noise vector b with $f(b) \propto exp(-\frac{\epsilon \cdot \|b\|_2}{2k} \cdot \frac{n_i \lambda}{q_{max} \cdot 2B})$. The noise vector b will be added to loss function as follows:

$$q_i \leftarrow \arg\min_{q_i} J_P(q_i, D_i) + b. \tag{10.15}$$

If q_i exceed the maximum value, normalized it by

$$q_i \leftarrow q_i \cdot \frac{q_{max}}{\|q_i\|_2}. \tag{10.16}$$

When finally obtain P and Q, the rating of u will be predicted by Eq. (10.7).

In general, Friedman et al. explored all possibilities on the perturbation of matrix factorization [82]. Similar to McSherry et al.'s method, they applied clamping to avoid exceeded noise on user-item matrix. But both methods are still suffered by high sensitivity. Even though they define a constant B to restrict the size of sensitivity, It is still too high comparing with the existing rating.

10.4 Private Neighborhood-Based Collaborative Filtering Method

Zhu et al. [258] mainly studied the trusted recommender systems. The algorithm ensured that a user cannot observe sensitive information from the recommendation outputs, and therefore, the proposed algorithm is immunized from a particular attack, *KNN attack*.

10.4.1 KNN Attack to Collaborative Filtering

Calandrino et al. [30] presented the *KNN attack*. They claim that if a recommendation algorithm and its parameters are known by an attacker, and supposing he/she knows the partial ratings history of active user u_a on m items, then the attacker can infer user u_a's remaining rating history. The inference process can be summarized as follows.

The attacker initially creates k fake users known as *sybils*. He/she arranges each *sybil*'s history rating with the m items in the active user u_a's rating history. Then with high probability, the k nearest neighbors of each *sybil* will consist of the other $k - 1$ *sybils* along with the active user u_a. The attacker inspects the lists of items recommended by the system to any of the sybils. Any item on the *sybils*, for example, lists not belonging to those m items, will be an item that u_a rates. The attacker will finally infer the ratings history of an active user and this process will be considered a serious privacy violation. While this is an example for *user-based* methods, similar inference can also be processed for item based methods.

A *KNN attack* can be performed efficiently in CF due to the sparsity of a typical rating dataset. Approximately, $m = O(\log n)$ is sufficient for an attacker to infer a user, where n is the total number of users in the rating dataset. For example, in a dataset with thousands of users, $m \approx 8$ is sufficient [30]. This is such a small number that can easily be collected by an attacker. Furthermore, an attack will be more serious if an attacker can adaptively change the rating history of his sybils by observing the output of CF. This can be easily implemented in a system that allows users to change previously entered ratings. How to hide similar neighbors is a major privacy issue that cannot be overlooked.

10.4.2   The Private Neighbor Collaborative Filtering Algorithm

For the privacy preserving issue in the context of *neighborhood-based* CF methods, the preserving targets differ between item-based methods and user-based methods due to the different perspectives regarding definition of similarity. In item-based methods, an adversary can infer who the neighboring users are by observing any changes in the item similarity matrix. Therefore, the objective is to protect the users' identity. In user-based methods, what an adversary can infer from the user similarity matrix is the item rated by the active user. The preserving objective is then to hide the historically rated items. The proposed *PNCF* algorithm can deal with both cases. To make it clear, Zhu et al. presented the *PNCF* algorithm from the perspective of the item-based methods, and this can be applied to user-based methods in a straightforward manner.

For traditional non-private item-based CF methods, the prediction of the active user's score is generated by previous ratings of this user on similar items. The first stage aims to identify the items of k nearest neighbor, and the second stage aims to predict the rating by aggregating the ratings on those identified neighbor items. To resist a *KNN attack*, the neighbor information in both stages should be preserved. The *PNCF* algorithm includes two private operations:

Private neighbor selection: prevents the adversary from inferring "who is the neighbor". Specifically, the *recommender mechanism* is adopted to perform private selection on the item similarity matrix to find k neighbors $N_k(t_i)$. The recommender mechanism ensures that for a particular item, deleting a user has little impact on the chosen probability. Therefore, the adversary is unlikely to figure out who their neighbors are by continuously observing recommendations, and is unlikely to infer the rating history of an active user by creating fake neighbors.

Predict perturbation: prevents the adversary from inferring "*what is the rating*" of a particular user on an item. It perturbs neighbors similarity by adding a zero mean Laplace noise to mask "*what is the rating given by a certain neighbor*". Noisy output is utilized as the weight in making predictions.

Figure 10.5 shows the private neighbor selection and perturbations in a neighborhood-based recommender system.

Algorithm 1 provides the pseudocode for how to use the proposed *PNCF* algorithm in a traditional *neighborhood-based* CF method. In the algorithm, Steps 1 and 4 are standard recommendations steps. Specifically, Step 1 computes the similarity between items. It is not a step to guarantee the privacy but will still play a vital role, because the result will be employed as a utility *score* in the next step. Step 4 provides the prediction for $\widehat{r}_{ai}$ according to Eq. (10.3). and considers it as an output of recommendations. However, Steps 2 and 3 implement two operations in the proposed *PNCF* algorithm. Compared to the *non-private* algorithm, the cost of this will be the prediction accuracy. The accuracy cost is defined as the utility loss of the algorithm.

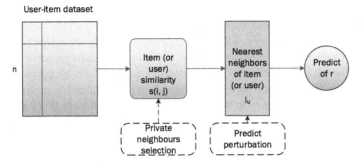

User-item dataset

Fig. 10.5 Different perturbation methods in neighbourhood based recommender system

---

Algorithm 1 Private Neighbor Collaborative Filtering($PNCF$)

Require: **D**, privacy parameter ϵ, truncated parameter w, number of neighbors k, u_a, t_i
Ensure: $\widehat{r}_{ai}$
    1. Computing item to item similarity Matrix S;
    2. Private Neighbor Selection: Select k neighbors $N_k(t_i)$ from I;
    3. Perturbation: Perturb the similarity in $N_k(t_i)$ by adding; $Lap(\frac{2k \cdot RS(i,\cdot)}{\epsilon})$ noise;
    4. Predict $\widehat{r}_{ai}$;

Although the standard differential privacy mechanism, the recommender mechanism, can be applied for private selection, it can not be directly applied to CF because the naive recommender mechanism induces abundant noise that significantly influences the performance of the prediction. Here, the main challenge is to enhance performance by decreasing the noise magnitude. Consequently, two issues will be addressed in Steps 2 and 3 accordingly: (a) decrease in the *sensitivity*, and (b) increased accuracy. Moreover, both of these operations consume an equivalent $\epsilon/2$. According to the sequential composition, the algorithm satisfies ϵ-differential privacy.

10.4.2.1 The Private Neighbor Selection

Private Neighbor Selection aims to privately select k neighbors from a list of candidates for the privacy preserving purpose. This is unlike the k nearest neighbor method which sorts all candidates by their similarities and selects the top k similar candidates. *Private Neighbor Selection* adopts the *recommender* mechanism to arrange probabilities for every candidate. The probability is measured by a score function and its corresponding *sensitivity*. Specifically, the similarity is used as the score function and the *sensitivity* is measured accordingly. For an item t_i, the score function q is defined as follows:

$$q_i(I, t_j) = s(i, j), \tag{10.17}$$

where $s(i, j)$ is the output of the score functions representing the similarity between t_i with t_j, I is item t_i's candidate list for neighbors, and t_j is the selected neighbor. The probability of selecting t_j will be arranged according to the $q_i(I, t_j)$.

The recommender mechanism uses the score function q to preserve differential privacy. However, the naive recommender mechanism fails to provide accurate predictions because it is too general, and therefore not suitable for recommendation purposes. Accordingly, two operations are proposed to address this obstacle. The first operation is to define a new *Recommendation-Aware Sensitivity* to decrease the noise, and the second operation is to provide a new recommender mechanism to enhance the accuracy. Both are integrated to form the proposed *PNCF* algorithm, which consequently obtains a better trade-off between privacy and utility.

10.4.2.2   Recommendation-Aware Sensitivity

This section presents *Recommendation-Aware Sensitivity* based on the notion of *Local Sensitivity* to reduce the magnitude of noise introduced for privacy-preserving purposes. *Recommendation-Aware Sensitivity* for score function q, $RS(i, j)$ is measured by the maximal change in similarity of two items when removing a user's rating record. Let $s'(i, j)$ denote the $s(i, j)$ after deleting a user, then $RS(i, j)$ captures the maximal difference if all the users' ratings are tested:

$$RS(i, j) = \max_{i,j \in I} ||s(i, j) - s'(i, j)||_1. \tag{10.18}$$

The result varies on the different item pairs.

Take the *item-based Cosin* similarity as an example to generate the value of *Recommendation-Aware Sensitivity*. PCC similarity can be measured in the same way. Items are considered as a vector in the n dimensional user space. For example, let $\mathbf{r_i}$ and $\mathbf{r_j}$ be a rating vector pair that contains all ratings given to t_i and t_j, respectively. Firstly, all the ratings for a single user u_x are selected to analyze his/her impact on $s(i, j)$. There are four possible rating pairs on both item t_i and t_j:

$$(0, 0), (0, r_{xj}), (r_{xi}, 0), \text{ and } (r_{xi}, r_{xj}).$$

Please note that in *neighborhood-based* CF methods, similarity is only measured on the co-rated set between two items. This means the first three rating pairs have no impact on the similarity function. Let $U_{ij} = \{u_x \in U | r_{xi} \neq \emptyset, r_{xj} \neq \emptyset\}$ be a set of users who rated both t_i and t_j, with these two item rating vectors then be represented as $\mathbf{r_{Ui}}$ and $\mathbf{r_{Uj}}$, respectively. The length of the vector, $||r_{Ui}||$, is determined by both the rating and the number of co-rated users. When deleting a user, the rating vector pair will be transferred to a new pair of $\mathbf{r'_{Ui}}$ and $\mathbf{r'_{Uj}}$ and the similarity will change to $s'(i, j)$ accordingly. The measurement and the smooth bound of *Recommendation-Aware Sensitivity* are summarized in Lemmas 10.1 and 10.2, respectively.

Lemma 10.1 *For any item pair t_i and t_j, the score function $q(I, t_j)$ has* Local Sensitivity

$$RS(i,j) = \max \left\{ \max_{u_x \in U_{ij}} \left(\frac{r_{xi} \cdot r_{xj}}{||r_i'|| \cdot ||r_j'||} \right), \max_{u_x \in U_{ij}} \left(\frac{r_{xi} \cdot r_{xj}(||r_i||||r_j|| - ||r_i||||r_j||)}{||r_i|| \cdot ||r_j|| \cdot ||r_i'|| \cdot ||r_j'||} \right) \right\},$$
(10.19)

where u_x is the user that makes a great impact on the t_i and t_j similarity, and $RS(i,j) = 1$ when $|U_{ij}| = 1$.

Proof

$$RS(i,j) = \max ||s(i,j) - s'(i,j)||_1$$
(10.20)

$$= \frac{r_i \cdot r_j}{||r_i|| \cdot ||r_j||} - \frac{r_i' \cdot r_j'}{||r_i'|| \cdot ||r_j'||}$$

$$= \frac{||r_i'|| \cdot ||r_j'|| \cdot r_i \cdot r_j - ||r_i|| \cdot ||r_j|| \cdot r_i' \cdot r_j'}{||r_i|| \cdot ||r_j|| \cdot ||r_i'|| \cdot ||r_j'||}.$$

Thus,

$$RS(i,j) \leq \begin{cases} \frac{||r_i|| \cdot ||r_i \cdot (r_i \cdot r_j - r_i' \cdot r_j')||}{||r_i|| \cdot ||r_j|| \cdot ||r_i'|| \cdot ||r_j'||}, & \text{if } ||r_i'|| \cdot ||r_j'|| \cdot r_i \cdot r_j \geq ||r_i|| \cdot ||r_j|| \cdot r_i' \cdot r_j', \\ \frac{r_i \cdot r_j(||r_i|| \cdot ||r_j|| - ||r_i'|| \cdot ||r_j'||)}{||r_i|| \cdot ||r_j|| \cdot ||r_i'|| \cdot ||r_j'||}, & \text{otherwise.} \end{cases}$$
(10.21)

Please note $\frac{||r_i|| \cdot ||r_i \cdot (r_i \cdot r_j - r_i' \cdot r_j')||}{||r_i|| \cdot ||r_j|| \cdot ||r_i'|| \cdot ||r_j'||} = \max_{x \in U_{ij}} \frac{r_{xi} \cdot r_{xj}}{||r_i'|| \cdot ||r_j'||}$.

Thus, $sim(i,j)$ and $sim'(i,j)$ differ at most by

$$\max \left\{ \max_{u_x \in U_{ij}} \left(\frac{r_{xi} \cdot r_{xj}}{||r_i'|| \cdot ||r_j'||} \right), \max_{u_x \in U_{ij}} \left(\frac{r_{xi} \cdot r_{xj}(||r_i||||r_j|| - ||r_i||||r_j||)}{||r_i|| \cdot ||r_j|| \cdot ||r_i'|| \cdot ||r_j'||} \right) \right\},$$
(10.22)

which is largely determined by the length of rating vector pairs.

Lemma 10.2 Recommendation-Aware Sensitivity *with smooth bound is*

$$B(RS(i,j)) = exp(-\beta) \cdot RS(i,j).$$
(10.23)

Both *Lemmas* 10.1 and 10.2 indicate that depending on the length of rating vector pairs, score function q will only change slightly in a normal case. Compared to *Global Sensitivity, Recommendation-Aware Sensitivity* can significantly decrease the noise magnitude. *Recommendation-Aware Sensitivity* is used in the *PNCF* algorithm. To simplify the notation, $RS(i,j)$ or *Recommendation-Aware Sensitivity* are used to represent $B(RS(i,j))$ in the following formulas.

10.4.2.3   Private Neighbor Selection Implementation

Private Neighbor Selection is the major private operation in *PNCF*. This section presents the *Private Neighbor Selection* based on the exponential mechanism and show how it is introduced into CF. Given an active user u_a and a target item t_i, a candidate item list I and a corresponding similarity list $\mathbf{s(i)} = < s(i, 1), \ldots, s(i, m) >$ are defined, which consists of similarities between t_i and other $m - 1$ items. According to Eq. (10.17), each element in $\mathbf{s(i)}$ is the score of the corresponding item. The goal of *Private Neighbor Selection* is to select k neighbors in candidate item list I, according to the score vector $\mathbf{s(i)}$. It should be noted that t_i is not selected in I as t_i is the target item.

However, even though *Recommendation-Aware Sensitivity* is applied to reduce noise, the naive exponential mechanism still yields low prediction accuracy. The reason is that performance of *neighborhood-based* CF methods is largely dependent on the quality of neighbors. If the top k nearest neighbors are assumed as the highest quality neighbors (s_k is used to denote the similarity to the k-th neighbor), the randomized selection process will have a high probability of picking up neighbors with low scores. The rating pattern of these low quality neighbors may be totally different from the pattern of the active user, which lowers the accuracy of the prediction. To address this problem, the best solution is to improve the quality of k neighbors under differential privacy constraints.

Motivated by this, the algorithm uses a new notion of *truncated similarity* as the score function to enhance the quality of selected neighbors. The truncated notion was first mentioned in Bhaskar et al.'s work [24], in which they used truncated frequency to decrease the computational complexity. The same notion can be used in our score function q to find those high quality neighbors. Specifically, for each item in the candidate list I, if its similarity $s(i, j)$ is smaller than $s_k(i, \cdot) - w$, then $s(i, j)$ is truncated to $s_k(i, \cdot) - w$, where w is a truncated parameter in the score function; otherwise it is still preserved as $s(i, j)$. The truncated similarity can be denoted as

$$\widehat{s}(i,j) = \max\left(s(i,j), s_k(i, \cdot) - w\right), \tag{10.24}$$

where truncated parameter w will be analyzed in the next section.

The *truncated similarity* ensures no item in $N_k(t_i)$ has a similarity less than $(s_k(i, \cdot) - w)$ and every item whose similarity is greater than $(s_k(i, \cdot) + w)$ is selected to the $N_k(t_i)$. Compared to the naive exponential mechanism in which some items with a similarity lower than $(s_k(i, \cdot) - w)$ may have a higher probability of being selected, the *truncated similarity* can significantly improve the quality of selected neighbors. Based on the *Recommendation-Aware Sensitivity* and the *truncated similarity*, the *Private Neighbor Selection* operation is presented in Algorithm 2.

Item candidate list I is divided into two sets: C_1 and C_0. Set C_1 consists of items whose similarities are larger than the truncated similarity $(s_k(i, \cdot) - w)$, and C_0 consists of the remaining items in I. In the *Private Neighbor Selection* operation, items in C_1 follow exponential distribution according to their similarities. Each

---

Algorithm 2 Private Neighbor Selection

Require: $\epsilon, k, w, t_i, I, \mathbf{s}(i)$
Ensure: $N_k(t_i)$
  1: Sort the vector $\mathbf{s}(i)$;
  2: $C_1 = [t_j | s(i,j) \geq s_k(i,j) - w, t_j \in I]$,
     $C_0 = [t_j | s(i,j) < s_k(i,j) - w, t_j \in I]$,
  3: **for** N=1:k **do**
  4:    **for each** item t_j in $\mathbf{t}_i$ **do**
  5:       Allocate probability as:

$$\frac{\exp\left(\frac{\epsilon \cdot \widehat{s(i,j)}}{4k \cdot RS(i,j)}\right)}{\sum_{j \in C_1} \exp\left(\frac{\epsilon \cdot s(i,j)}{4k \cdot RS(i,j)}\right) + |C_0| \cdot \exp\left(\frac{\epsilon \cdot \widehat{s(i,j)}}{4k \cdot RS(i,j)}\right)}.$$

  6:    **end for**
  7:    Sample an elements t from C_1 and C_0 without replacement according to their probability;
  8:    $N_k(t_i) = N_k(t_i) + t$;
  9: **end for**

---

of these have the probability proportion to $\exp\left(\frac{\epsilon \cdot s(i,j)}{4k \cdot RS(i,j)}\right)$. In C_0, the algorithm does not deal with the elements one by one. Instead, it considers C_0 as a single candidate with a probability proportion to $\exp\left(\frac{\epsilon \cdot \widehat{s(i,j)}}{4k \cdot RS(i,j)}\right)$. When C_0 is chosen, the items in C_0 are selected uniformly. This does not violate differential privacy because C_0 is used as a single candidate and assign the weight according to the exponential mechanism. This means the probability for each element in C_0 still follows exponential distribution. The probability of providing the same output for neighboring datasets is still bounded by $exp(\epsilon)$.

The *Private Neighbor Selection* operation can provide high quality neighbors and guarantee the differential privacy simultaneously. The choice of w will influence the utility in a fixed privacy level.

10.4.3   Privacy and Utility Analysis

10.4.3.1   Utility Analysis

The utility of *PNCF* is measured by the accuracy of the predictions. Given an input dataset D, the *non-private* neighbourhood-based CF method is set as a baseline. By comparing the selected neighbors in *PNCF* with the corresponding neighbors in the baseline method, the utility level of the proposed algorithms can be analyzed.

To predict the rating of t_i for the active user u_a, the *non-private* algorithm typically chooses the top k similar neighbors and then generates an integrated output as the prediction. Therefore, the closeness between top k similar neighbors in the baseline method and the privately selected neighbors in *PNCF* is the key factor that determines the utility level. When implementing the exponential mechanism, the randomized selection step will choose low similarity neighbors with high probability, which significantly lowers the accuracy of recommendations.

The proposed *PNCF* algorithm can preserve the quality of the neighbors in two aspects: every neighbor with a true similarity greater than $(s_k + w)$ will be selected; and no neighbor in the output has a true similarity less than $(s_k - w)$, where the true similarity refers to the original similarity without perturbation or truncation. Therefore, *Private Neighbor Selection* guarantees that with high probability, the k selected neighbors will be close to the actual top k nearest ones. Hence, the utility in *PNCF* will be better retained than that in the naive exponential mechanism. Two theorems and proofs are used to support the claims as follows.

Theorem 10.1 *Given an item t_i, let $N_k(t_i)$ be the k selected neighbors and $|v|$ be the maximal length of all the rating vector pairs. Suppose RS as the maximal* Recommendation-Aware Sensitivity *between t_s and other items respectively, and suppose ρ is a small constant less than 1. Then, for all $\rho > 0$, with probability at least $1 - \rho$, the similarity of all the items in $N_k(t_i)$ are larger than $s_k - w$, where* $w = \min(s_k, \frac{4k \cdot RS}{\epsilon} \ln \frac{k \cdot (|v|-k)}{\rho})$.

Proof First, the algorithm computes the probability of selecting a neighbor with a similarity less than $(s_k - w)$ in each round of sampling. This occurs when there is an unsampled neighbor with similarity no less than s_k. Then with the constraint of w and ρ, no neighbor in the output has true similarity less than $(s_k - w)$ after k rounds sampling for neighbor selection.

If a neighbor with similarity $(s_k - w)$ is still waiting for selection, the probability of picking a neighbor with similarity less than $(s_k - w)$ is $\leq \frac{\exp(\frac{\epsilon(s_k-w)}{4k \cdot RS})}{\exp(\frac{\epsilon s_k}{4k \cdot RS})} = \exp(-\frac{\epsilon \cdot w}{4k \cdot RS})$. Since there are at most $|v|$ neighbors with similarity less than $(s_k - w)$, according to the union bound, the probability of choosing a neighbor with similarity less than $(s_k - w)$ is at most $(|v| - k) \cdot \exp(-\frac{\epsilon \cdot w}{4k \cdot RS})$.

Furthermore, by the union bound in the sampling step, the probability of choosing any neighbor with similarity less than $(s_k - w)$ is at most $k \cdot (|v| - k) \cdot \exp(-\frac{\epsilon \cdot w}{4k})$.

Let $\rho \geq k \cdot (|v| - k) \cdot exp(-\frac{\epsilon \cdot w}{4k \cdot RS})$.

Then,

$$-\frac{w \cdot \epsilon}{4k \cdot RS} \leq \ln\left(\frac{\rho}{k \cdot (|v|-k)}\right) \tag{10.25}$$

$$\Rightarrow \quad \frac{w \cdot \epsilon}{4k \cdot RS} \geq \ln \frac{k \cdot (|v|-k)}{\rho}$$

$$\Rightarrow \quad w \geq \frac{4k \cdot RS}{\epsilon} \ln \frac{k \cdot (|v|-k)}{\rho}.$$

Thus, the probability that similarities less than $(s_k - w)$ will be chosen is less than ρ. As defined in Sect. 10.4.2.2, *Recommendation-Aware Sensitivity* is $O(\frac{1}{||v||^2})$.

So for constant ρ, $s = O(\frac{k \cdot \ln(|v|-k)}{\epsilon ||v||^2})$ is sufficient. In practise, the algorithm has to ensure $s_k - w \geq 0$ in COS similarity or $s_k - w > -1$ in PCC, so $w = \min(s_k, \frac{4k \cdot RS}{\epsilon} \ln \frac{k \cdot (|v|-k)}{\rho})$.

Theorem 10.2 *Given an item t_i, for all $\rho > 0$, with probability at least $1 - \rho$, the similarities of all neighbors $> s_k + w$ are present in $N_k(t_i)$, where $w = \min(s_k, \frac{4k \cdot RS}{\epsilon} \ln \frac{k \cdot (|v|-k)}{\rho})$.*

Proof Similar to Theorem 10.1, the algorithm firstly computes the probability of picking a neighbor with a similarity less than s_k in each round of sampling when an unsampled neighbor with similarity greater or equal than $(s_k + w)$ is not present in $N_k(t_i)$. Then we prove that with the constraint of w and ρ, all neighbors with similarity $\geq s_k + w$ have been chosen in $N_k(t_i)$.

Suppose a neighbor with a similarity greater than $(s_k + w)$ has not been selected in $N_k(t_i)$, then the conditional probability of picking any neighbor with similarity less than s_k is $\leq \frac{\exp(\frac{\epsilon(s_k)}{4k \cdot RS})}{\exp(\frac{\epsilon(s_k+w)}{4k \cdot RS})} = \exp(-\frac{\epsilon \cdot w}{4k})$. Therefore, the probability of not selecting any neighbor with similarity less than s_k in any of the k rounds of sampling is:

$$\left(1 - (|v| - k) \cdot \exp(\frac{\epsilon(s_k + w)}{4k \cdot RS})\right)^k \geq \left(1 - k \cdot (|v| - k) \cdot \exp(\frac{\epsilon(s_k + w)}{4k \cdot RS})\right).$$
$$(10.26)$$

Let $1 - \rho \leq (1 - k \cdot (|v| - k) \cdot \exp(\frac{\epsilon(s_k+w)}{4k \cdot RS}))$. Thus, similarly to the proof of Theorem 10.1. When

$$w = \min\left(s_k, \frac{4k \cdot RS}{\epsilon} \ln \frac{k \cdot (|v| - k)}{\rho}\right), \qquad (10.27)$$

all neighbors with similarity $\leq s_k + w$ are present in $N_k(t_i)$.

10.4.3.2   Privacy Analysis

The proposed *PNCF* algorithm contains two private operations: the *Private Neighbor Selection* and the *Perturbation*. The *Private Neighbor Selection* is essentially processing the exponential mechanism successively. An item is selected without replacement in each round until k distinct neighbors are chosen. The score *sensitivity* is calibrated by *Recommendation-Aware Sensitivity*.

From the definition of the exponential mechanism, each selection round preserves $(\frac{\epsilon}{2k})$-differential privacy. The *sequential composition* undertakes the privacy guarantee for a sequence of differentially private computations. When a series of private analysis is performed sequentially on a dataset, the *privacy budget* ϵ will be added for each step. According to the *sequential composition* definition, *Private Neighbor Selection* guarantees $\frac{\epsilon}{2}$-differential privacy as a whole.

The *Perturbation* step adds independent Laplace noise to the $N_k(t_i)$ chosen in the previous step. Given an item set $N_k(t_i)$, perturbation adds independent Laplace noise to their similarities. The noise is calibrated by $\epsilon/2$ and the *Recommendation-Aware Sensitivity* is:

$$s_{noise}(i,j) = s(i,j) + Lap\left(\frac{2 \cdot RS(i,j)}{\epsilon}\right).$$ (10.28)

According to the definition of the Laplace mechanism, this step satisfies $\epsilon/2$-differential privacy.

Consequently, when combining both operations, the proposed method preserves ϵ-differential privacy by applying *composition lemma* on the selection and perturbation step together.

10.4.4   Experiment Analysis

10.4.4.1   Datasets and Measurements

The datasets are the popular `Netflix` dataset[1] and the `MovieLens` dataset.[2] The `Netflix` dataset was extracted from the *Netflix* Prize dataset, where each user rated at least 20 movies, and each movie was rated by 20–250 users. The `MovieLens` dataset includes around one million ratings collected from 6040 users about 3900 movies. `MovieLens` is the standard benchmark data for collaborative filtering research, while the `Netflix` dataset is a real industrial dataset released by Netflix. Both datasets contain millions of ratings that last for several years and are sufficient for investigating the performance of the proposed method from both a research and industry perspective. Specifically, the *All-But-One* strategy is applied, which randomly selects one rating of each user, and then, predicts its value using all the left ratings in the dataset.

To measure the quality of recommendations, a popular measurement metric is applied, *Mean Absolute Error* (MAE) [7]:

$$MAE = \frac{1}{|T|} \sum_{a,i \in T} |r_{ai} - \widehat{r}_{ai}|,$$ (10.29)

where r_{ai} is the true rating of user u_a on item t_i, and $\widehat{r}_{ai}$ is the value of predicting the rating. T denotes the test dataset, and $|T|$ represents its size. A lower *MAE* means a higher prediction accuracy. In each experiment, the traditional *non-private* CF method is used as a baseline.

---

[1]http://www.netflixprize.com.

[2]http://www.grouplens.org.

10.4.4.2   Performance of *PNCF*

This section, examines the performance of *PNCF* on two similarity measurement metrics, PCC and COS. Specifically, the experiment applied the traditional *neighborhood-based* CF as the *non-private* baseline, and then compared the *PNCF* with the standard DP method in terms of the recommendation accuracy, as both quantify the privacy risk to individuals. Parameter k represent the number of the neighbors. Moreover, the truncated parameter w was set according to Lemma 10.1. The privacy budget ϵ was fixed to 1 to ensure the *PNCF* algorithm satisfies the *1-differential privacy*.

Table 10.3 shows the results on the *Netflix* data set. From this table, it is clear that *PNCF* significantly outperforms DP in all configurations. Specifically, in the *item-based* manner with the PCC metric, when $k = 40$, *PNCF* achieves a MAE of 0.7178. This outperforms DP by 13.60%. When $k = 10$, *PNCF* obtains a MAE of 0.7533, which outperforms DP by 13.07%. In the *user-based* manner with the PCC metric, when $k = 30$, *PNCF* outperforms DP by 8.71% in MAE. Similar trends are also observed when measuring neighbor similarity in COS. This indicates that *PNCF* performs better in terms of the recommendation accuracy than the standard differential privacy (DP) method that uses *Global Sensitivity* and the naive

Table 10.3 Overall performance comparison on *Netflix*

	k	PCC			COS		
		Non-private	PNCF	DP	Non-private	PNCF	DP
Item-based	5	0.7504	0.7835	0.9153	0.7524	0.7893	0.8786
	10	0.7210	0.7533	0.8666	0.7240	0.7637	0.8301
	15	0.7121	0.7407	0.8499	0.7159	0.7486	0.8134
	20	0.7083	0.7343	0.8371	0.7137	0.7414	0.8012
	25	0.7070	0.7278	0.8329	0.7133	0.7401	0.7972
	30	0.7068	0.7244	0.8287	0.7137	0.7390	0.7927
	35	0.7072	0.7208	0.8244	0.7152	0.7385	0.7899
	40	0.7078	0.7178	0.8154	0.7169	0.7398	0.7835
	45	0.7086	0.7163	0.8062	0.7185	0.7392	0.7789
	50	0.7092	0.7146	0.8019	0.7199	0.7395	0.7770
User-based	5	0.7934	0.8025	0.8962	0.8041	0.8009	0.8661
	10	0.7641	0.7691	0.8538	0.7691	0.7708	0.8237
	15	0.7509	0.7551	0.8324	0.7553	0.7612	0.8042
	20	0.7428	0.7485	0.8229	0.7481	0.7564	0.7956
	25	0.7375	0.7435	0.8137	0.7434	0.7525	0.7867
	30	0.7339	0.7408	0.8115	0.7398	0.7512	0.7847
	35	0.7316	0.7398	0.8081	0.7371	0.7494	0.7812
	40	0.7298	0.7381	0.8059	0.7350	0.7480	0.7794
	45	0.7284	0.7368	0.8028	0.7331	0.7469	0.7764
	50	0.7273	0.7353	0.8015	0.7317	0.7456	0.7751

Table 10.4 Paired-*t*-test for *PNCF* vs. DP on *Netflix*

		df	t	p-value
Item-based	PCC	9	26.5135	<0.0001
	COS	9	11.6696	<0.0001
User-based	PCC	9	25.8884	<0.0001
	COS	9	10.5045	<0.0001

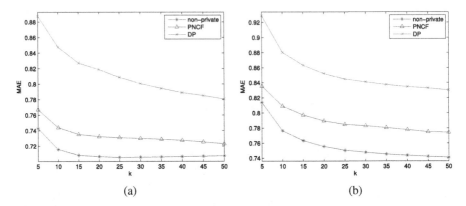

Fig. 10.6 Performance of *PNCF* and DP on the *MovieLens* dataset. (**a**) PCC-item. (**b**) PCC-user

exponential mechanism. On the other hand, compared with the baseline *non-private* algorithm, the accuracy cost introduced by *PNCF* is much smaller than the cost introduced by DP. This is because *PNCF* introduces two novel operations to reduce the magnitude of introduced noise. Moreover, the two-tailed, paired *t*-test with a 95% confidence level has been applied to evaluate the performance of *PNCF* under all configurations. The detailed statistical results on the *Netflix* dataset are presented in Table 10.4. This table shows that the difference in performance between *PNCF* and DP is statistically significant.

Moreover, Fig. 10.6 shows the results on the *MovieLens* dataset. Specifically, Fig. 10.6a, b shows the performance of *PNCF* and DP with the PCC metric under an *item-based* and *user-based* manner, respectively. It is clear that *PNCF* performs much better then DP. In addition, it is observed that as k increases, both *PNCF* and DP achieve better MAE. However, *PNCF* always performs better than DP across all k values. This is because *PNCF* achieves privacy preserving by distinguishing the quality of potential neighbors, and therefore always selects good-quality neighbors as analysed in Sect. 10.4.3.1. Moreover, *PNCF*'s MAE performance is very close to that of the *non-private* baseline. This indicates *PNCF* can retain the accuracy of recommendation while providing comprehensive privacy for individuals.

10.5 Summary

One of the most popular recommendation techniques is the Collaborative Filtering (CF) method, which predicts the rating on an unknown item based on the ratings on that item by other similar users or neighbors. However, an adversary can infer the rating history of an active user by creating fake neighbors based on some background information. After a brief discussion on the differentially private untrustworthy recommender system and the differentially private trustworthy recommender system, this chapter focuses on the privacy preserving issue in the context of neighbourhood-based CF methods. The following three research questions are addressed in this chapter to design the private neighborhood-based collaborative filtering (PNCF) method: (1) How to preserve neighborhood privacy? (2) How to define sensitivity for recommendation purposes? And (3) how to design the differentially private recommender mechanism? The design of the PNCF method considers all possible privacy leakage and integrates with the sensitivity and mechanism requirement of applications. It uses a novel recommendation-aware sensitivity to reduce the large magnitude of noise found in other methods and uses a private neighbor selection mechanism to protect neighbors.

Chapter 11
Privacy Preserving for Tagging Recommender Systems

11.1 Introduction

The widespread success of social network web sites, such as *Del.icio.us* and *Bibsonomy*, introduces a new concept called the *tagging recommender system* [107]. These social network web sites usually enable users to annotate resources with customized tags, which in turn facilitates the recommendation of resources. Over the last few years, a large collection of data has been generated, but the issue of privacy in the recommender process has generally been overlooked. An adversary with background information may re-identify a particular user in a tagging dataset and obtain the user's historical tagging records [84]. Moreover, in comparison with traditional recommender systems, tagging recommender systems involve more semantic information that directly discloses users' preferences. Hence, the privacy violation involved is more serious than traditional violations [177]. Consequently, how to preserve privacy in tagging recommender systems is an emerging issue that needs to be addressed.

Over the last decade, a variety of privacy preserving approaches have been proposed for traditional recommender systems [178]. For example, cryptography is used in the rating data for multi-party data sharing [31, 247]. Perturbation adds noise to the users' ratings before rating prediction [181, 182], and *obfuscation* replaces a certain percentage of ratings with random values [23]. However, these approaches can hardly be applied in tagging recommender systems due to the semantic property of tags. Specifically, cryptography completely erases the semantic meaning of tags, while perturbation and obfuscation can only be applied to numerical values instead of words. These deficiencies render these approaches impractical in tagging recommendation. To overcome these deficiencies, the tag suppression method has recently been proposed to protect a user's privacy by modeling users' profiles and eliminating selected sensitive tags [177]. However, this method only releases an incomplete dataset that significantly affects the recommendation performance.

© Springer International Publishing AG 2017   131
T. Zhu et al., *Differential Privacy and Applications*,
Advances in Information Security 69, DOI 10.1007/978-3-319-62004-6_11

This chapter introduces differential privacy into tagging recommender systems, with the aim of preventing re-identification of users and avoiding the association of sensitive tags (e.g., healthcare tags) with a particular user. However, although these characteristics make differential privacy a promising method for tagging recommendation, there remains some barriers to research:

- The basic differential privacy mechanism only focuses on releasing statistical information that can barely retain the structure of the tagging dataset. For example, this naive mechanism lists all the tags, counts the number and adds noise to the statistical output, but ignores the relationship among users, resources and tags. This simple statistical information is inadequate for recommendations.
- Differential privacy utilizes the randomized mechanism to preserve privacy, and usually introduces a large amount of noise due to the sparsity of the tagging dataset. For a dataset with millions of tags, the randomized mechanism will result in a large magnitude of noise.

Both barriers imply the basic differential privacy mechanism cannot be simply applied in a tagging recommender system, and a novel differentially private mechanism is needed. To overcome the first barrier, a synthetic dataset can be generated to retain the relationship among tags, resources and users rather than releasing simple statistical information. The second barrier can be addressed by shrinking the randomized domain, because the noise can decrease when the randomized range is limited. For example, the topic model method is a possible way to structure tags into groups and limit the randomized domain within each topic. Based on these observations, this chapter proposes a tailored differential privacy mechanism that optimizes the performance of recommendation with a fixed level of privacy. Table 11.1 shows the basic setting of tagging recommender system application.

Table 11.1 Application settings

Application	Tagging recommender systems
Input data	User-item tagging dataset
Output data	Synthetic tagging dataset
Publishing setting	Non-interactive
Challenges	Data sparsity
Solutions	Shrink domain by using topic model
Selected mechanism	Dataset partitioning
Utility measurement	Error measurement
Utility analysis	Marlkov inequality
Privacy analysis	Parallel composition
Experimental evaluation	Performance measured by semantic loss

11.2 Preliminaries

11.2.1 Notations

In a tagging recommender system, D is a tagging dataset consisting of users, resources and tags. Let $U = \{u_1, u_2, \cdots\}$ be a set of users, $R = \{r_1, r_2, \cdots\}$ be a set of resources, and $T = \{t_1, t_2, \cdots\}$ be the set of all tags, the relationships among users, resources and tags are defined as *folksonomy* $F =< U, R, T, AS >$, where the ternary relationship $AS \subseteq U \times R \times T$ is referred to as the *tag assignment* set. For a particular user $u_a \in U$ and a resource $r_b \in R$, the authors use $T(u_a, r_b)$ to represent all tags flagged by the u_a on r_b, and use $T(u_a)$ to denote all tags utilized by user u_a. The recommended tags for u_a on a given resource $r \in R$ are represented by $T'(u_a, r)$, When a user u_a select a particular resource r, the system will recommend a suitable tag to the user. To achieve the target, the system first generates a rank on a set of tags according to some quality or relevance criteria, then the top-N ranked tags, $\widehat{T}(u_a, r)$ are finally selected as recommended tags.

Tagging dataset D can be structured by a set of users' profiles $\mathbf{P} = \{P(u_1), \ldots, P(u_{|U|})\}$. A user u_a's profile $P(u_a) =< T(u_a), W(u_a) >$ is usually modeled by his tagging records, including tag's names $T(u_a) = \{t_1, \ldots, t_{|T(u_a)|}\}$ and weights $W(u_a) = \{w_1, \ldots, w_{|T(u_a)|}\}$ [177].

11.2.2 Tagging Recommender Systems

A considerable amount of literature has explored various techniques for tagging recommendations, which offers users the possibility to annotate resources with personalized tags and to ease the process of finding suitable tags for a resource [201]. For example, *Del.icio.us*[1] allows the sharing of bookmarks; *Bibsonomy*[2] is a website sharing of bibliographic references; *Last.fm*[3] and *Netflix*[4] allow users to tag on music and movies, respectively. These tagging recommendations have similar functions: a user can add a resource and assign arbitrary tags to it. The collection of all his assignments constitute the folksonomy. The user can explore his collections, as well as the collections of other users.

To achieve the goal of tagging recommendation. Sigurbjornsson et al. provided a typical recommender strategy [203]. Given a resource with user-defined tags, an ordered candidate list of candidate tags is derived for each user-defined tag based on tag co-occurrence. After aggregating and ranking in the candidate list, the system provides top-N ranked tags.

---

[1]https://del.icio.us/.
[2]http://www.bibsonomy.org/.
[3]http://www.last.fm/.
[4]https://www.netflix.com/.

Another well-known study is *FolkRank* [107], which adapts the *PageRank* method into the tagging recommender system. The key idea of *FolkRank* is that a resource flagged with important tags by important users becomes important itself. The importance is measured by weight $\vec{\omega}$, which is computed iteratively as follows.

$$\vec{\omega} \leftarrow \lambda A \vec{\omega} + (1 - \lambda) \vec{\rho}, \tag{11.1}$$

where A is the *adjacency matrix* of folksonomy, $\vec{\rho}$ is the preference vector and $\lambda \in [0, 1]$ is the damping factor measuring the influence of $\vec{\rho}$.

There are other methods for tagging recommender systems, such as the clustering based method [201], the tensor decomposition [212], and the topic-model method [130].

11.2.3 Related Work

Compared to general recommender systems, the privacy problem in tagging recommendation systems is more complicated due to its unique structure and semantic content. Parra-Arnau et al. [177] made the first contribution towards the development of a privacy preserving tagging system by proposing the suppression approach. They first modeled the user's profile using a tagging histogram and eliminated sensitive tags from this profile. To retain utility, they applied a clustering method to structure all tags and to suppress the less represented ones. Finally, they analyzed the effectiveness of their approach by discussing the semantic loss of users. However, there are several limitations on tag suppression. It only releases an incomplete dataset, with parts of the sensitive tags deleted, and sensitive tags are subjective without any quantity measurement. Furthermore, if the dataset is publicly shared, users can be identified because the remaining tags still have the potential to reveal a user's identity. The privacy issue in tagging recommender systems remains largely unexplored.

Zhu [256] proposed a differentially private tagging release algorithm, with the aim of preserving comprehensive privacy for individuals and maximizing the utility of the synthetic published dataset.

11.3 Private Tagging Publishing Method

11.3.1 User Profiles

In tagging dataset D, a user's profile is defined by $P(u) = < T(u), W(u) >$, in which tags in are represented by $T(u) = \{t_1, \ldots, t_{|T|}\}$, and weights are denoted as $W(u_a) = \{w_1, \ldots, w_{|T|}\}$, where $w_i = 0$ indicates that t_i is unused. A set of users

Fig. 11.1 User profile

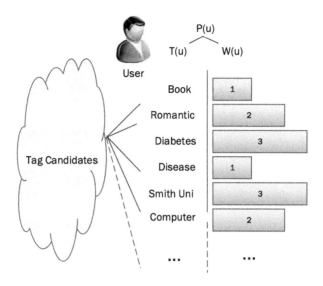

in D is presented by $\mathbf{P} = \{P(u_1), \ldots, P(u_{|U|})\}$. Figure 11.1 shows an example of a user's profile, which has tags and related weights. All tags are derived from a tag candidate set. User's profile $P(u)$ may disclose the user's privacy. If an adversary has part of the information on $P(u)$, he/she may re-identify a particular user in a tagging dataset by simply searching the known tags. More background information results in higher probability of re-identifying a user.

If we apply differential privacy mechanism directly, for a user u_a, noise will be added to the weight $W(u_a)$ and a noisy profile $\widehat{P}(u_a) =< T(u_a), \widehat{W}(u_a) >$ will be published. However, in this case, $W(u_a)$ is a sparse vector because a user tends to flag limited tags. When applying the randomized mechanism, $\widehat{W}(u_a)$ will contain a large amount of noise because lots of weights in $W(u_a)$ will change from zero to a positive value.

One way to reduce the noise is to shrink the randomized domain, which refers to the diminished number of zero weights in the profile. To achieve this objective, the authors structure the tags into K topics and each user is represented by a topic-based profile $P_z(u_a) =< T_z(u_a), W_z(u_a) >$, where $T_z(u_a) = \{T_{z_1}(u_a), \ldots, T_{z_K}(u_a)\}$ represents tags in each topic and $W_z(u_a) = w_{z_1}(u_a), \ldots, w_{z_K}(u_a)$ is the frequency of tags in a topic z. Compared to $W(u_a)$, $W_z(u_a)$ is less sparse. Because the noise added to each $w_{z_i} \in W_z(u_a)$ is equal to $w_i \in W(u_a)$, the total noise added to $W_z(u_a)$ will significantly diminish. Figure 11.2 shows an example of topic based profile, in which all tags are from a particular topic, and related weight refers to the occurrence of a topic.

In the follow section, a **Private Topic-based Tagging Release** (PriTop) algorithm is proposed to address the privacy issues in tagging recommender systems. The authors in [256] first present an overview of the algorithm, then provide details of its operations.

Fig. 11.2 User topic based
profile

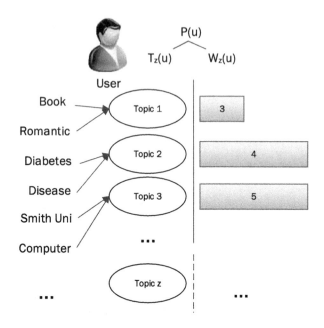

11.3.2 *Private Tagging Release Algorithm Overview*

The *PriTop* algorithm aims to publish all users' profiles by masking their exact tags
and weights under the notion of differential privacy. Three private operations are
introduced to ensure each user in the released dataset cannot be re-identified.

Private Topic Model Generation   This creates multiple topics according to the
    resources and tags by masking the topic distribution on tags. From the output,
    the adversary cannot infer to which topic a tag belongs.
Topic Weight Perturbation   This operation masks the weights of tags in a user's
    profile to prevent an adversary from inferring how many tags a user has annotated
    on a certain topic.
Private Tag Selection   Some privately selected tags replace the original tags

On the basis of these private operations, the proposed *PriTop* algorithm generates
a new tagging dataset for recommendations, and its pseudocode is provided in
Algorithm 1. Firstly, step 1 divides the privacy budget into three parts for three
private operations. Step 2 groups all tags into K topics and in step 3, the weight
for each topic is perturbed by Laplace noise. After privately selecting the new tags
to replace the original ones in step 4, the sanitized dataset $\widehat{D}$ is finally released for
recommendation purposes in the last step.

The *PriTop* algorithm contains three private operations: *Private Topic Model
Generation*, *Topic Weight Perturbation* and *Private Tag Selection*. The privacy
budget ϵ is consequently divided into three pieces, as illustrated in Table 11.2.

Algorithm 1 Private Topic-Based Tagging Release (PriTop) Algorithm

---

Require: D, privacy parameter ϵ, K.

Ensure: $\widehat{D}$

    1. Divided privacy budget into $\epsilon/2$, $\epsilon/4$ and $\epsilon/4$;

    2. *Private Topic Generation*: create topic-based user profiles $P(u_a)$ based on the private topic model with $\epsilon/2$ privacy budget;

for each user u_a **do**

    3. *Topic Weight Perturbation*: add Laplace noise to the topic weights with $\epsilon/4$ privacy budget;

$$\widehat{W}(u_a) = W(u_a) + Laplace(\frac{4}{\epsilon})^K.$$

    for each topic z_k in $P(u_a)$ **do**

        4. *Private Tag Selection*: Select tags according to the $\widehat{W}(u_a)$ in $\epsilon/4$ privacy budget;

    end for

end for

    5. Output $\widehat{D}$ for tagging recommendations;

Table 11.2 Privacy budget allocation in *PriTop* algorithm

Operations	Privacy budget
Private topic model generation	$\epsilon/2$
Topic weight perturbation	$\epsilon/4$
Private tag selection	$\epsilon/4$

These three private operations simultaneously guarantee a fixed ϵ-differential privacy and retain the acceptable recommendation performance. Details for the Private Topic Model Generation operation is presented in following sections.

11.3.3 Private Topic Model Generation

This operation categorizes unstructured tags into topics to eliminate the randomization domain. The method is based upon the idea of a topic model, which considers a document as a mixture of topics and a topic is a probability distribution over words [25]. *Latent Dirichlet Allocation* (LDA) is the simplest topic model and the intuition behind it is that documents exhibit multiple topics [26]. Moreover, it is concise, clear and easy to be extended in terms of privacy preserving. Therefore, we applied *LDA* as the baseline model and introduced differential privacy to generate a private topic model. In this model, a resource is considered as a document and a tag is interpreted as a word. After discovering a set of topics expressed by tags, *Laplace mechanism* ensures that an adversary cannot infer to which topic a tag belongs.

We conceptualize the *Private Topic Model Generation* in three steps:

- Generating LDA model using the *Gibbs Sampling* approach [87].
- Adding Laplace noise to the LDA model to create a private model.
- Creating user's profile according to the private LDA model.

11.3.3.1 LDA Model Construction

The first step constructs the LDA model by *Gibbs Sampling*. LDA makes the assumption that there are K topics associated with a given set of documents, and that each resource in this collection is composed of a weighted distribution of these topics. Let $Z = \{z_1, \ldots z_K\}$ be a group of topics, with the following equation representing a standard LDA model to specify the distribution over tag t.

$$Pr(t|r) = \sum_{l=1}^{K} Pr(t|z_l)Pr(z_l|r), \qquad (11.2)$$

where $Pr(t|z_l)$ is the probability of tag t under a topic z_l and $Pr(z_l|r)$ is the probability of sampling a tag from topic z in the resource r.

In the model, the main variables of interest are the topic-tag distribution $Pr(t|z)$ and the resource-topic distribution $Pr(z|r)$. *Gibbs sampling* is a relatively efficient method to estimate these variables in the model by extracting a set of topics from a large documents collection [87]. It iterates multiple times over each tag t of resource r and samples the new topic z for the tag based on the posterior probability $Pr(z|t_i, r, Z_{-i})$ by Eq. (11.3) until the model converges.

$$Pr(z|t_i, r, Z_{-i}) \propto \frac{C_{tK}^{TK} + \beta}{\sum_{t_i}^{|T|} C_{t_iK}^{TK} + |T|\beta} \frac{C_{rk}^{RK} + \alpha}{\sum_{K=1}^{K} C_{r_iK}^{RK} + K\alpha}, \qquad (11.3)$$

where C^{TK} maintains a count of all topic-tag assignments and C^{RK} counts the resource-topic assignments. Z_{-i} represents all topic-tag assignment and resource-topic assignment except the current z for t_i. α and β are hyperparameters for the Dirichlet priors, which can be interpreted as the prior observation for the counts.

Evaluation on the $Pr(t|z)$ and $Pr(z|r)$ is formulated as follows:

$$Pr(t|z) = \frac{C_{tk}^{TK} + \beta}{\sum_{t_i}^{|T|} C_{t_iK}^{TK} + |T|\beta}. \qquad (11.4)$$

$$Pr(z|r) = \frac{C_{rK}^{RK} + \alpha}{\sum_{k=1}^{K} C_{r_iK}^{RK} + K\alpha}. \qquad (11.5)$$

After converging, the LDA model is generated with estimating $Pr(z|t, r)$, $P(t|z)$ and $P(z|r)$.

11.3.3.2 Private Model Generation

The second step introduces differential privacy by adding Laplace noise to the final counts in the LDA model. There are four difference counts in Eq. (11.3):

C_{tK}^{TK}, $\sum_{t_i}^{|T|} C_{t_iK}^{TK}$, C_{rK}^{RK} and $\sum_{K=1}^{K} C_{r_iK}^{RK}$. If we change the topic assignment on current t_i, the C_{tK}^{TK} will decrease by 1 and $\sum_{t_i}^{|T|} C_{t_iK}^{TK}$ will increase by 1. Similarly, if the C_{rK}^{RK} decreases by 1, the $\sum_{K=1}^{K} C_{r_iK}^{RK}$ will increase by 1 accordingly. Based on this observation, we sample two groups of independent random noise from Laplace distribution and add them to four count parameters. The new $\widehat{Pr}(z|t,r)$ is evaluated as follows:

$$\widehat{Pr}(z|t,r) \propto \frac{C_{tK}^{TK} + \eta_1 + \beta}{\sum_{t_i}^{|T|} C_{t_iK}^{TK} - \eta_1 + |T|\beta} \frac{C_{rK}^{RK} + \eta_2 + \alpha}{\sum_{K=1}^{K} C_{r_iK}^{RK} - \eta_2 + K\alpha}, \tag{11.6}$$

where η_1 and η_2 are both sampled from Laplace distribution $Laplace(\frac{2}{\epsilon})$ with the *sensitivity* as 1.

11.3.3.3   Topic-Based Profile Generation

The third step creates topic-based user profiles. For each user with tags $T(u_a) = \{t_1, \ldots, t_{|T(u_a)|}\}$ and related resources $R(u_a) = \{r_1, \ldots, r_{|R(u_a)|}\}$, each tag can be assigned to a particular topic $z_l \in Z$ according to the $\widehat{Pr}(z|t,r)$. So the user profile can be represented by a topic-based

$$P_z(u_a) = < T_z(u_a), W_z(u_a) >$$

with the weight $W_z(u_a) = \{w_1, \ldots, w_K\}$.

Generally, *Private Topic Model Generation* operation constructs a private LDA model to create the topic-based user profile and retain the structure between tags and resources. Details of this operation are shown in Algorithm 2.

For resources, the resulting topics represent a collaborative view of the resource and tags of topics reflect the vocabulary to describe the resource. For users, the resulting topics indicate the preference of a user.

11.3.4   Topic Weight Perturbation

After generating the topic-based user profile $P_z(u_a)$, Laplace noise will be added to mask the counts of tags in each topic.

$$\widehat{W}_z(u_a) = \{\widehat{w}_{z_1}(u_a), \ldots, \widehat{w}_{z_K}(u_a)\} + Laplace\left(\frac{4}{\epsilon}\right)^K. \tag{11.7}$$

Noise added on the weight $W_z(u_a)$ implies the revision of the tag list $T_z(u_a)$. Positive noise indicates that new tags are added to the $T_z(u_a)$, while negative noise

---

Algorithm 2 *Private Topic Generation*

Require: D, privacy budget $\frac{\epsilon}{2}$, numbers of topics K
Ensure: $\mathbf{P_z} = \{P_z(u_1), \ldots P_z(u_{|U|})\}$
  for each tag t_i **do**
    Randomly initial the topic assignment $Pr(z = z_j | t_i, r, Z_{-i})$;
  end for
  for each r_j **do**
    for each t_i **do**
      repeat
        estimate $Pr(t|z) = \frac{C_{tk}^{TK} + \beta}{\sum_{t_i}^{|T|} C_{t_ij}^{TK} + |T|\beta}$;
        estimate $Pr(z|r) = \frac{C_{rk}^{RK} + \alpha}{\sum_{k=1}^{K} C_{rik}^{RK} + K\alpha}$;
        resign t_i a new topic z according to

$$Pr(z|t_i, r, Z_{-i}) \propto \frac{C_{tk}^{TK} + \beta}{\sum_{t_i}^{|T|} C_{t_iK}^{TK} + |T|\beta} \frac{C_{rk}^{RK} + \alpha}{\sum_{k=1}^{K} C_{rik}^{RK} + K\alpha}.$$

      until converge
    end for
  end for
  for each r_j **do**
    for each t_i **do**
      Sample noise from the Laplace distribution:

$$\eta_1 \sim Laplace(\frac{2}{\epsilon}),$$

$$\eta_2 \sim Laplace(\frac{2}{\epsilon}).$$

      Estimate $\widehat{Pr}(z|t, r) \propto \frac{C_{tK}^{TK} + \eta_1 + \beta}{\sum_{t_i}^{|T|} C_{t_iK}^{TK} - \eta_1 + |T|\beta} \frac{C_{rK}^{RK} + \eta_2 + \alpha}{\sum_{k=1}^{K} C_{rik}^{RK} - \eta_2 + K\alpha}$;
    end for
  end for
  for each user u_a **do**
    for each $< r, t >$ pair **do**
      Assign z_i according to $Pr_{noise}(z|t, r)$;
      Generate $P_z(u_a)$;
    end for
  end for
  Output $\mathbf{P_z} = \{P_z(u_1), \ldots P_z(u_{|U|})\}$.

---

indicates some tags have been deleted from the list. For positive noise in the topic z_l, the operation will choose the tags with the highest probability in the current topic z_j according to the $Pr(t|z)$. For negative noise, the operation will delete the tag with the lowest probability in the current topic z_j. The adding and deleting operations will be defined by Eqs. (11.8) and (11.9)

$$\widetilde{T}_{z_l}(u_a) = T_{z_l}(u_a) + t_{new}, \tag{11.8}$$

where $t_{new} = \max_{i=1}^{|T|} Pr(t_i|z_l)$.

$$\widetilde{T}_{z_l}(u_a) = T_{z_l}(u_a) - t_{delete}, \qquad (11.9)$$

where $t_{delete} = \min_{i=1}^{|T|} Pr(t_i|z_l)$.

After perturbation, we use $\widetilde{P}_z(u_a) =< \widetilde{T}_z(u_a), \widehat{W}_z(u_a) >$ to represent the noisy topic-based user profile. However, the $\widetilde{P}_z(u_a)$ still has the high probability to be re-identified because it retains a major part of the original tags. The next operation will replace all tags in $\widetilde{T}(u_a)$ to preserve privacy.

11.3.5 Private Tag Selection

The *Private Tag Selection* operation manages to replace original tags with selected new tags. The challenge is how to select suitable new tags in related topics. For a tag $t_i \in \widetilde{T}_{z_l}(u_a)$, uniformly random tag selection within $\widetilde{T}_{z_l}(u_a)$ is unacceptable due to significant utility detriment. The intuitive solution to retain utility is to use the most similar tag to replace the original one. However, this approach is also dangerous because the adversary can easily figure out the tag most similar using simple statistical analysis. Consequently, the *Private Tag Selection* needs to: (1) retain the utility of tags, and (2) mask the similarities between tags.

To achieve these, *Private Tag Selection* adopts the exponential mechanism to privately select tags from a list of candidates. Specifically, for a particular tag t_i, the operation first locates the topic z_l to which it belongs and all tags in $\widehat{T}_{z_l}(u_a)$ are then included in a candidate list I. Each tag in I is associated with a probability based on a *score function* and the *sensitivity* of the function. The selection of tags is performed based on the allocated probabilities.

The *score function* is defined by the distance between tags. In the LDA model, the distance between tags is measured by the extent of the same shared topics [208]. Using a probabilistic approach, the distance between two tags t_1 and t_2 is computed based on the *Jensen-shannon divergence* (*JS divergence*) between $Pr(z|t_i = t_1)$ and $Pr(z|t_i = t_2)$. *JS* divergence is a symmetrized and smoothed version of the *Kullback-Leibler divergence* (*KL divergence*). Pr_1 and Pr_2 is defined to be:

$$D_{KL}(Pr_1||Pr_2) = \sum_i \ln\left(\frac{Pr_1(i)}{Pr_2(i)}\right) Pr_1(i), \qquad (11.10)$$

$$D_{JS}(Pr_1||Pr_2) = \frac{1}{2}D_{KL}(Pr_1||S) + \frac{1}{2}D_{KL}(Pr_2||S), \qquad (11.11)$$

where $S = \frac{1}{2}(Pr_1 + Pr_2)$.

Algorithm 3 Private Tag Selection

Require: $\frac{\epsilon}{4}, \widetilde{T}_z(u_a), Pr(z|t)$

Ensure: $\widehat{T}_z(u_a)$
    for each tags t_i in $T_z(u_a)$ **do**
        1. located the t_i in topic z_l;
        for each tags t_j in z_l **do**
            2. Allocate probability as:

$$\frac{\exp\left(\frac{\epsilon \cdot q_i(I, t_j)}{8}\right)}{\sum_{j \in z_l} \exp\left(\frac{\epsilon \cdot q_i(I, t_j)}{8}\right)};$$

    end for
        3. Select a tag t_j from z_l without replacement according to the probability;
    end for
    4. Output $\widehat{T}_z(u_a)$

Because the JS divergence is bounded by 1 when using the base 2 logarithm [147], so we define the score function q for a target tag t_i as follows:

$$q_i(I, t_j) = (1 - D_{JS}(Pr_i||Pr_j), \tag{11.12}$$

where I is tag t_i's candidate list, and $t_j \in I$ are the candidate tags for replacement. Each tag t_j has a score according to Eq. (11.12).

The *sensitivity* for score function q is measured by the maximal change in the distance of two tags when removing a topic shared by both t_i and t_j. Let $D'_{JS}(Pr_i||Pr_j)$ denote the new distance between t_i and t_j after deleting a topic, and the maximal difference between $D'_{JS}(Pr_i||Pr_j)$ and $D_{JS}(Pr_i||Pr_j)$ is bounded by 1. Therefore, the sensitivity of score function is 1.

On the basis of the *score function* and *sensitivity*, the probability arranged to each tag t_j is computed by Eq. (11.13) with the privacy budget $\frac{\epsilon}{4}$. The pseudocode of *Private Tag Selection* is presented in Algorithm 3:

$$\frac{\exp\left(\frac{\epsilon \cdot q_i(I, t_j)}{8}\right)}{\sum_{j \in z_l} \exp\left(\frac{\epsilon \cdot q_i(I, t_j)}{8}\right)}, \tag{11.13}$$

where z_l is the topic in which t_j belongs to.

11.3.6 Privacy and Utility Analysis

The proposed *PriTop* aims to obtain the acceptable utility with a fixed ϵ differentially privacy level. This section first proves the algorithm is satisfied with the ϵ-differential privacy, then analyzes the utility cost.

11.3.6.1 Privacy Analysis

To analyze the privacy guarantee, we apply two composite properties of the privacy budget: the sequential and the *parallel composition*. Based on the above Lemmas and privacy budget allocation in Table 11.2, we measure the privacy level of our algorithm as follows:

- The *Private Topic Model Generation* operation is performed on the whole dataset with the privacy budget $\frac{\epsilon}{2}$. This operation preserves $\frac{\epsilon}{2}$-differential privacy.
- The *Topic Weight Perturbation* applies the Laplace mechanism to the weights of topics. The noise is calibrated by $Lap\left(\frac{4}{|T(u)|\cdot\epsilon}\right)^K$ and preserves $\frac{\epsilon}{4}$ − *differentialprivacy* for each user. Furthermore, as a user's profile is independent, replacing a user's tags has no effect on other user profiles. The *Private Tag Selection* preserves $\frac{\epsilon}{4}$-differential privacy as a whole.
- The *Private Tag Selection* processes the exponential mechanism successively. For one user u, each tag in the profile is replaced by a privately selected tag until all tags are replaced. Each selection is performed on the individual tags, therefore the selection for each user guarantees $\frac{\epsilon}{4}$-differential privacy. Similar to the previous operation, every user can be considered as subsets of the entire dataset. Thus, the *Private Tag Selection* guarantees $\frac{\epsilon}{4}$-differential privacy.

  Consequently, the proposed *PriTop* algorithm preserves ϵ-differential privacy.

11.3.6.2 Utility Analysis

Given a target user u_a, the utility level of the proposed *PriTop* algorithm is determined by the accuracy of the tagging recommendation, which is highly dependent on the distance between $P(u_a)$ and $\widehat{P}(u_a)$ [177]. The distance between $P(u_a)$ and $\widehat{P}(u_a)$ is referred to as *semantic loss* [177].

$$SLoss = \frac{1}{|U|}\sum_{u\in U}\left(\frac{\sum_{t\in P(u_a)} d(t,\widehat{t})}{\max d \cdot |T(u)|}\right), \tag{11.14}$$

where $\widehat{t}$ is the new tag replacing the tag t.

If we consider each private step as query f, then the difference between $f(D)$ and $f(D)$ is the *sematic loss*. We then apply a widely used error utility definition. We will demonstrate the *sematic loss* is bounded by a certain value α with a high probability.

All three private steps affect the *semantic loss*. But the first step, private topic model generation, only affects the distance measurement between tags. Therefore, we only need to measure the $SLoss_1$ in the perturbation step and $SLoss_2$ in the selection step.

Theorem 3 *For any user $u \in U$, for all $\delta > 0$, with probability at least $1 - \delta$, the $SLoss_1$ of the user in the perturbation is less than α. When*

$$|T(u)| \geq \frac{K \cdot \exp(\frac{-\epsilon\alpha_a}{4})}{\delta},$$

the perturbation operation is satisfied with (α, δ)-useful.

Proof The perturbation adds Laplace noise with $\epsilon/4$ privacy budge to the weight of each topic in a user's profile. According to the property of Laplace distribution $Lap(b)$:

$$Pr(|\gamma| > t) = Pr(\gamma > t) + Pr(\gamma < t) = 2 \int_t^\infty x \exp\left(-\frac{x}{b}\right) dx = \quad (11.15)$$

$$= exp\left(-\frac{t}{b}\right). \quad (11.16)$$

We have

$$Pr(SLoss_1 > \alpha_a) = \frac{2K \cdot d\left(t_{ai}, \widehat{t}_{ai}\right)}{\max d|T(u_a)|} \int_{\alpha_a}^0 \frac{\epsilon}{8} \exp\left(-\frac{\epsilon x}{4}\right) dx = \quad (11.17)$$

$$Pr(SLoss_1 > \alpha_a) = \frac{K \cdot d\left(t_{ai}, \widehat{t}_{ai}\right)}{\max d|T(u_a)|} \exp\left(-\frac{\epsilon\alpha_a}{4}\right). \quad (11.18)$$

As the perturbation step adds new tags or delete tags, the $d(t_{ai}, \widehat{t}_{ai})$ will be less than the maximal value. When we use the *JS* divergence, the maximal $d(t_{ai}, \widehat{t}_{ai})$ is 1, so we obtain the evaluation on the $SLoss_1$ is

$$Pr(SLoss_1 < \alpha_a) \leq 1 - \frac{K \cdot \exp\left(-\frac{\epsilon\alpha_a}{4}\right)}{|T(u_a)|}. \quad (11.19)$$

Let

$$1 - \frac{K \cdot \exp\left(-\frac{\epsilon\alpha_a}{4}\right)}{|T(u_a)|} \geq 1 - \delta,$$

thus

$$|T(u_a)| \geq \frac{K \cdot \exp\left(\frac{-\epsilon\alpha_a}{4}\right)}{\delta}. \tag{11.20}$$

The average semantic loss for all the users is less than the maximal value, $\alpha = \max_{u_a \in U} \alpha_a$, we have

$$|T(u)| \geq \frac{K \cdot \exp\left(\frac{-\epsilon\alpha_a}{4}\right)}{\delta}. \tag{11.21}$$

Theorem 3 reveals the *semantic loss* of perturbation depends on the number of tags a user has. More tags results in a lower *semantic loss*.

Theorem 4 *For any user $u \in U$, for all $\delta > 0$, with probability at least $1 - \delta$, the SLoss$_2$ of the user in the private selection is less than α. When*

$$Q \leq \frac{\exp(\frac{\epsilon}{8})}{1 - \delta\alpha},$$

where Q is the normalization factor that depends on the topic that $t \in T(u)$ belongs to, the private selection operation is satisfied with (α, δ)-useful.

Proof According to Marlkov's inequality, we obtain

$$Pr(SLoss_2 > \alpha_a) \leq \frac{E(SLoss_2)}{\alpha_a}. \tag{11.22}$$

For each tag t_{ai} in $\widehat{P}_a$, the probability of 'unchange' in the private selection is proportional to $\frac{\exp(\frac{\epsilon}{8})}{Q_i}$, where Q_i is the normalization factor depending on the topic t_{ai} belongs to. Therefore, we obtain

$$E(SLoss_2) = \sum_{t_i \in T(u_a)} \frac{d(t_{ai}, \widehat{t}_{ai})}{\max d|T(u_a)|} \left(1 - \frac{\exp\left(\frac{\epsilon}{8}\right)}{Q_i}\right).$$

According to (11.22), the evaluation of the SLoss$_2$ is

$$Pr(SLoss_2 > \alpha_a) \leq \frac{\sum_{t_i \in T(u_a)} d(t_{ai}, \widehat{t}_{ai}) \left(1 - \frac{\exp(\frac{\epsilon}{8})}{Q_i}\right)}{|T(u_a)|\alpha_a}.$$

When we take the maximal $d(t_{ai}, \widehat{t}_{ai})$ and $Q = \max Q_i$, it can be simplified as

$$Pr(SLoss_2 \leq \alpha_a) \geq 1 - \frac{1 - \frac{1}{Q} \exp\left(\frac{\epsilon}{8}\right)}{\alpha_a}. \tag{11.23}$$

Let

$$1 - \frac{1 - \frac{1}{Q}\exp\left(\frac{\epsilon}{8}\right)}{\alpha_a} \geq 1 - \delta,$$

thus

$$Q \leq \frac{\exp\left(\frac{\epsilon}{8}\right)}{1 - \delta\alpha_a}. \tag{11.24}$$

For all users, α is determined by the maximal value: $\alpha = \max_{u_a \in U} \alpha_a$.
Finally, we obtain

$$Q \leq \frac{\exp\left(\frac{\epsilon}{8}\right)}{1 - \delta\alpha}, \tag{11.25}$$

where $Q = \max Q_i$, and $Q_i = \sum_{j \in z_l} \exp\left(\frac{\epsilon \cdot d(t_i, t_j)}{8}\right)$.

The proof shows the *semantic loss* of private selection mainly depends on the privacy budget and the normalization factor Q_i, which is measured by the total distance inside topic z to which t_i belongs. The shorter distance leads to a smaller Q_i and less *semantic loss*.

Further analysis shows that the total distance in a topic is determined by the privacy budget ϵ in the *private topic model generation*. It can be concluded that the privacy budget has significant impact on the utility level of *PriTop*.

11.3.7 Experimental Evaluation

11.3.7.1 Datasets

We conduct the experiment on four datasets: *Del.icio.us, Bibsonomy, MovieLens* and *Last.fm*. All of them are collected from collaborative tagging systems, which allow users to upload their resources, and to label them with arbitrary words. The statistics for all datasets are summarized in Table 11.3. *Del.icio.us* and *Bibsonomy* datasets focus on resources and tags sharing, so each user tends to collect more resources and various of tags. *Last.fm* and *MovieLens* datasets are derived from traditional recommender systems, comparing with particular tagging systems, they have less various of tags and less number of tags on each resources. To demonstrate the effectiveness of the proposed *PriTop*, we select datasets from both tagging systems and traditional recommender systems.

All four datasets are structured in the form of triples (*user, resource, tag*), and filtered by automatically removing redundant tags like "imported", "public", etc.

Table 11.3 Characteristics of the datasets

Dataset	Record	$|U|$	$|R|$	$|T|$
Del.icio.us	130,160	3000	34,212	12,183
Bibsonomy	163,510	3000	421,928	93,756
Last.fm	186,479	1892	12,523	9749
MovieLens	47,957	2113	5,908	9079

Del.icio.us dataset   is retrieved from the *Del.icio.us* web site by the *Distributed Artificial Intelligence Laboratory* (DAI-Labor),[5] and includes around 132 million resources and 950,000 users. We extracted a subset with 3000 users, 34,212 bookmarks and 12,183 tags.

Bibsonomy dataset   is provided by *Discovery Challenge 2009 ECML/PKDD2009.*[6] The dataset contains 3000 individual users, 421,928 resources and 93,756 tags.

MovieLens and *Last.fm* datasets   both were obtained from *HetRec 2011,*[7] which were generated by the *Information Retrieval Group* at *Universidad Autonoma de Madrid*.

11.3.7.2 Performance of Tagging Recommendation

This section investigates the effectiveness of *PriTop* in the context of tagging recommendations and compares it with *tag suppression*. We apply a state-of-the-art tagging recommender system, *FolkRank* [107], to measure the degradation of tag recommendations with privacy preserving.

In the *FolkRank* configuration, we follow [107] to apply a typical setting with $\lambda = 0.7$, $\overrightarrow{p} = 1$, and the preference weights are set to $1 + |U|$ and $1 + |R|$, respectively. The computation repeats for ten iterations or stops when the distance between two consecutive weight vectors is less than 10^{-6}. Please note that the choice of parameters on *FolkRank* is less important because the target of the experiment is to evaluate the impact of the private operations rather than recommendation performance.

We apply the *Leave-One-Out* measurement strategy, which is a popular configuration in evaluating tag recommendations [153]. To begin with, we randomly select one resource of each user, and predict a list of N (top-N list) tags using all remaining tags in the dataset. *Precision* and *recall* are used to quantify the performance. A large value of *precision* or *recall* means better performance.

---

[5]http://www.dai-labor.de/.

[6]http://www.kde.cs.uni-kassel.de/ws/dc09/.

[7]http://ir.ii.uam.es/hetrec2011.

$$precision(T(u,r), T'(u,r)) = \frac{T(u,r) \cap \widetilde{T}(u,r)}{|T'(u,r)|}. \qquad (11.26)$$

$$recall(T(u,r), T'(u,r)) = \frac{T(u,r) \cap T'(u,r)}{|T(u,r)|}. \qquad (11.27)$$

The following experiments compare the *PriTop* with *tag suppression* when N varies from 1 to 10. For *PriTop*, we chose the number of topic $K = 100$, and test the performance when $\epsilon = 1$ and $\epsilon = 0.5$. For *tag suppression*, we fix the *eliminate parameter* to $\sigma = 0.8$ and $\sigma = 0.6$, which corresponds to the suppression rates of 0.2 and 0.4, respectively.

Figure 11.3 presents the recall of recommendation results. It is observed that the proposed *PriTop* algorithm significantly outperforms the *tag suppression* method on both privacy budgets. Specifically, as shown in Fig. 11.3a, when $N = 1$, *PriTop* achieves a *recall* at 0.0704 with the $\epsilon = 1$ which outperforms the result from the *tag suppression* with $\sigma = 0.6$, 0.0407, by 42.19%. This trend is retained as the increasing of N. For example, when $N = 5$, *PriTop* achieves a *recall* at 0.1799 with the $\epsilon = 1$ which outperforms the result from the *tag suppression* by 37.19% when $\sigma = 0.6, 0.113$. When N reaches 10, the *PriTop* still retains 36.09% higher on *recall* than *tag suppression*. Even we choose the lower privacy budget with $\epsilon = 0.5$ and a higher *eliminate* parameter *sigma* $= 0.8$, the improvement of *PriTop* is still significant. The *PriTop* has a *recall* of 0.1382, which is also 7.67% higher than *tag suppression* with a *recall* of 0.1276. Moreover, the improvement of *PriTop* is more obvious when $N = 10$. It achieves *recalls* of 0.1882 and 0.2408 when $\epsilon = 1$ and $\epsilon = 0.5$, respectively. But *tag suppression* only achieves *recalls* of 0.1538 and 0.1881 with $\sigma = 0.6$ and $\sigma = 0.8$. Similar trends can also be observed in Fig. 11.3b–d. For example, in the *MovieLens* dataset, when $N = 10$ and $\epsilon = 1.0$, the recall of *PriTop* is 0.4445, which is 27.33% higher than *tag suppression* with $\sigma = 0.8$. With the same configuration, *PriTop* is 22.43% and 25.22% higher than *tag suppression* in *Last.fm* and *Bibsonomy* datasets. The experimental results show the *PriTop* algorithm outperforms *tag suppression* in variety of N, which implies that *PriTop* can retain more useful information for recommendations than simply deleting the tags.

Moreover, it is clear that the performance of *PriTop* is very close to the *non-private* baseline. For example, in Fig. 11.3a, when $\epsilon = 1$, the *recall* of the *De.licio.us* dataset is 0.2408, which is only 3.00% lower than the non-private recommender result. Other datasets show the same trend. As shown in Fig. 11.3b–d, with the same configuration, the *PriTop* result is 3.62% lower than the non-private result in the *MovieLens* dataset, 7.58% lower in the *Last.fm* dataset and 1.4% lower in the *Bibsonomy* dataset. The results indicate the *PriTop* algorithm can achieve the privacy preserving objective while retaining a high accuracy of recommendations.

Figure 11.4 supports the above claims by plotting the precision results, which also shows the improvement of *PriTop* compared to *tag suppression* and the high recommender accuracy results of *PriTop*. However, curves in Fig. 11.3d are not as smooth as others. This may be caused by the statistical property of *Bibsonomy*

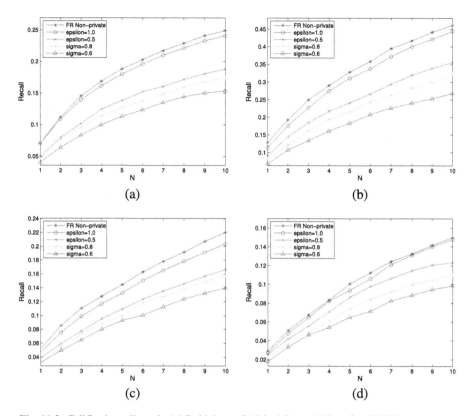

Fig. 11.3 *FolkRank* recall result. (**a**) Del.icio.us. (**b**) MovieLens. (**c**) Last.fm. (**d**) Bibsonomy

dataset, which contains a large number of tags that appear only once. When the test samples include a large proportion of these tags, the *precision* fluctuates.

11.4 Summary

In comparison with traditional recommender systems, tagging recommender systems involve more semantic information that directly discloses users' preferences, therefore, the potential privacy violation involved is more serious than traditional violations. Consequently, how to preserve privacy in tagging recommender systems is an emerging issue that needs to be addressed. Simply apply the naive differential privacy mechanism cannot achieve the desired privacy-preserving goals for a tagging recommender system. This chapter presented a private topic-based tagging release (PriTop) algorithm to address the privacy issues in tagging recommender systems. The algorithm generates a synthetic dataset that retains the relationship among tags, resources and users rather than releasing simple statistical information,

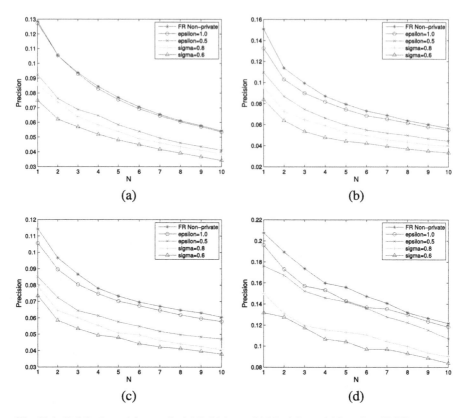

Fig. 11.4 *FolkRank* precision result. (**a**) Del.icio.us. (**b**) MovieLens. (**c**) Last.fm. (**d**) Bibsonomy

and shrinks the randomized domain decrease noise. The PriTop algorithm can publish all users' profiles by masking their exact tags and weights under the notion of differential privacy, and a novel private topic model-based method is used to structure the tags and to shrink the randomized domain.

Current evaluation only concentrates on one recommender algorithm, *FolkRank*, with other recommendation techniques, such as the tensor decompositions method, requiring further investigations. Also we would like to explicitly explore the semantic similarity in tags to help improve the tradeoff of privacy and utility.

Chapter 12
Differentially Location Privacy

12.1 Introduction

The advances in sensor-enable devices, such as mobiles and wearable techniques have allowed for the location information to be available in social media. The popularity of location-based services has resulted in a wealth of data on the movements of individuals and populations. However, the places that people visit will disclose extremely sensitive information about their behaviours, home and work locations, preferences and habits. Location privacy is an emerging issue that needs to be addressed.

Although location information brings added value for personalizing the consumers' experience, it is widely believed that such uses of location data are invasions of privacy. People may feel reluctant to disclose their locations to others because of privacy concerns. Through analysing these data, it is possible to infer an individual's home location, political views and religious inclinations, etc. Moreover, if location databases are abused by authorized users or broken by intruders, adversary can then attempt security attacks. Accordingly, the ability to preserve the privacy of users as the location information is involved is an essential requirement for social networks and LBSs.

Differential privacy applies the randomized mechanism to add controlled noise into numeric or non-numeric values, and has been proven effective in sensitive data release. Recently, a number of works have attempted to bring differential privacy into location data release [36], though three challenges with this strategy still remains open:

- The first challenge occurs when differential privacy utilizes the randomized mechanism to preserve privacy, it usually introduces a large amount of noise due to the sparsity of the dataset. For a dataset with millions of locations, the randomized mechanism will result in a large magnitude of noise.

© Springer International Publishing AG 2017

T. Zhu et al., *Differential Privacy and Applications*,
Advances in Information Security 69, DOI 10.1007/978-3-319-62004-6_12

Table 12.1 Application settings

Application	Location based service
Input data	Trajectory
Output data	Synthetic trajectory
Publishing setting	Non-interactive
Challenges	High sensitivity; data sparsity
Solutions	Adjust sensitivity measurement, using clustering to shrink the domain
Selected mechanism	Dataset partitioning
Utility measurement	Error measurement
Utility analysis	Union bound
Privacy analysis	Parallel composition
Experimental evaluation	Performance measured by distance error

- The second challenge lies in the measurement of sensitivity. For a location dataset, the randomization mechanism is associated with distance measurements. However, if we just measure the sensitivity by the traditional way, which involves the maximal distance between locations, the sensitivity will be very large. To achieve a rigorous privacy guarantee, a large magnitude of noise has to be added, and this will significantly decrease the utility of the location dataset.
- The third challenge is the semantic retaining of the location data. When randomizing a location dataset, the traditional differential privacy mechanism does not consider the semantic of the locations. Typically, only the distance-based measure was used to perturb the location regardless of which `town/city/country` it belongs to. For example, the locations in `Niagara Falls` (on the border of `USA` and `Canada`) could likely be perturbed into a different country if not considering the semantics.

All these challenges imply that differential privacy should not be adopted to location dataset in a straight forward manner. Previous work [36] can only solve parts of these challenges. A novel mechanism is in high demand. This chapter proposed two basic methods, Geo-indistinguishability [36] and synthesization method [97]. After that, a hierarchical location publishing method [242] is proposed in detail. Table 12.1 shows the application settings.

12.2   Preliminary

Let D be a location dataset as shown in Table 12.2. Each record contains a user with all the locations where he/she visited. U is the set containing all the users, and $|U|$ is the number of the users. A location set is denoted as $\mathscr{X}$ and each location is a point $x \in \mathscr{X}$.

Table 12.2 Location dataset

User	Locations
u_1	$< x_{11}, y_{11} >, < x_{12}, y_{12} >, \ldots$
u_2	$< x_{21}, y_{21} >, < x_{22}, y_{22} >, \ldots$
$\ldots$	$\ldots$
$u_{|U|}$	$< x_{n1}, y_{n1} >, < x_{n2}, y_{n2} >, \ldots$

$T(u_a)$ is the set of locations where u_a has ever been, and $|T(u_a)|$ is the number of those locations. A query f is a function that maps the data set D to an abstract range $\mathfrak{R}$: $f : D \to \mathfrak{R}$. We typically consider of $\mathfrak{R}$ as the set of possible outputs of a mechanism performed on a dataset. For example, If f is a count query, the abstract range will fall into a real number domain R. A group query is represent by F.

Table 12.2 shows an example of location dataset, in which each user may travels to different locations and a location can appear several times for the same user. Apparently, the statistical information on a location dataset, say, a histogram indicating the frequency of each location, is potentially valuable for location knowledge discovering. In data analysis or mining mechanism, aggregate results are abstracted from the statistical information to response the queries such as *"which locations are the most attractive?"* and *"how many users requested LBS at an attractive location?"*

To achieve the privacy preserving, answers to the queries should be obfuscated. The goal of this work is to propose an efficient method for location data release with differential privacy while maintaining adequate utility.

12.3 Basic Location Privacy Methods

There are two possible queries for a location-based service: snapshot and continuous queries [207]. A snapshot query is a request submitted once by the user. For example, *Where is the closest Sushi bar?* A continuous query is submitted at discrete time points by the same user. For example, *Continuously send me gas price coupons as I travel the interstate highway.* Both types of queries are prevalent nowadays in location based systems. Location privacy preserving ensures that from these queries, adversary cannot infer the current location (from snapshot query) or the trajectory (from continuous query) of the user. For location-based services, location privacy calls for methods that preserve as much as the quality of the desired services, while hindering the undesired tracking capacities of those services.

12.3.1 Snapshot Location Privacy: Geo-Indistinguishability

Some types of snapshot queries can use existing differential privacy mechanisms. For example, when a location based server would like to hide the number of people in a particular region, a typical range query publishing mechanism can be used. Cormode [49] applied spatial decomposition methods, which is a type of dataset partitioning mechanism, to decrease noise. They instantiated a hierarchical tree structure to decompose a geometric space into smaller areas with data points partitioned among the leaves. Noise is added to the count for each node. Similarly, Zhang et al. [253] applied spatial decomposition methods in the problem of private location recommendations, which is the extension of range queries. Quadtree is used to partition the region and noise is added to protect users' historical trajectories.

Some location based applications need to hide the exact locations of individuals. Chatzikokolakis et al. [36] proposed a new notion, *geo-indistinguishability*, which protects an individual's exact location, while disclosing enough location information to obtain the desired service. Its main idea relates to a differential privacy level of the radius that the individual has visited: for any radius $r > 0$, an individual will have (ϵ, r)-privacy.

Most aforementioned work are concerned with the applications involving aggregate location information about several people. For the protection of a single person's location, Dewri [56] incorporated differential privacy to k-anonymity by fixing an anonymity set of k locations. The proposed method requires that the probability of outputting the same obfuscated location from any set of anonymized k locations should be similar. This property is achieved by adding Laplace noise to each coordinate independently.

Geo-indistinguishability considered a query to provider for nearby Sushi bar in a private way [36], i.e., by disclosing some approximate information z instead of his exact location x. The method solve the problem of what level of privacy guarantee can the user expect in this scenario? They considered the level of privacy within a radius r. A user has a ℓ-privacy with r, in which ℓ represent user's privacy level proportional to radius. The smaller ℓ is, the higher the privacy. Therefore, Andrés et al. define geo-indistinguishability as follows: A mechanism satisfies ϵ-geo-indistinguishability iff for any radius $r > 0$, the user has ϵr-privacy within r. This definition shows that a user is protected within any radius r, but with a level $\ell = \epsilon r$ with the distance. Suppose ϵ is fixed, when $r = 1\,\mathrm{km}$, ℓ is small, which means the user enjoys a higher privacy level. Figure 12.1 illustrates the idea of privacy levels decreasing with the radius r.

The key idea of geo-indistinguishability is to make two points in radius r indistinguishable, which is defined by a probabilistic model.

Fig. 12.1
Geo-indistinguishability
privacy level

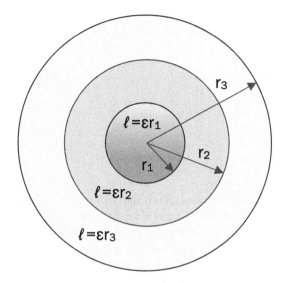

12.3.1.1 Probabilistic Model

Suppose there are $\mathscr{X}$ of point of interest, which might be user's possible locations, $\mathscr{Z}$ is set as user's reported values, which can be arbitrary. Assuming user has visited $x \in \mathscr{X}$, but report $z \in \mathscr{Z}$. The attacker's background information is modeled by a prior distribution π on $\mathscr{X}$, where $\pi(x)$ is the probability assigned to location x. As the perturbed location z can be obtained by adding random noise to the actual location x, z can be considered as a probabilistic value. The mechanism K is a probabilistic function assigning each location a probability distribution on $\mathscr{Z}$. For example, $K(x)(Z)$ is the probability of reported point of x in $Z \subseteq \mathscr{Z}$. Each observation $Z \in \mathscr{Z}$ of a mechanism K induces a posterior distribution $\sigma = Bayes(\pi, K, Z)$ on K, defined as

$$\sigma(x) = \frac{K(x)(Z)\pi(x)}{\sum_{x'} K(x')(Z)\pi(x')}. \tag{12.1}$$

Multiplicative distance is defined to measure the distance between two distribution on some set $\mathscr{S}$:

$$d_p(\sigma_1, \sigma_2) = sup_{S \in \mathscr{S}} \left| ln \frac{\sigma_1(S)}{\sigma_2(S)} \right|. \tag{12.2}$$

12.3.1.2 Geo-Indistinguishability Definition

Based on the probabilistic model, the geo-indistinguishability guarantees that for any x and x', whose Euclidean distance $d(x, x')$ is less than r, the distance between

Fig. 12.2
Geo-indistinguishability
definition

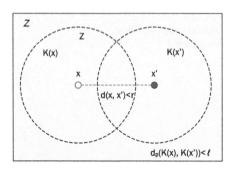

two distributions, $K(x)$ and $K(x')$, should be less than ℓ. Figure 12.2 illustrate definition. When x and x' are fixed, two circles are used to represent distributions of $K(x)$ and $K(x')$ in point set Z. The distance of $K(x)$ and $K(x')$ should be bounded in the privacy level ℓ. When meeting with two set of points tuples $\mathbf{x_1} = x_0, \ldots$ and $\mathbf{x_2} = x_0, \ldots$, the distance between two tuples are defined by the maximum distance between two points in these two tuples.

Definition 12.1 (Geo-Indistinguishability)   A mechanism K satisfies ϵ-geo-indistinguishability iff for all x, x':

$$d_p(K(x), K(x')) \leq \epsilon d(x, x').$$ (12.3)

According to Eq. (12.2), the definition can be formulated to

$$K(x)(Z) \leq e^e dx, x' K(x')(Z)$$ (12.4)

for all $x, x' \in \mathscr{X}$, and $Z \subseteq \mathscr{Z}$.

12.3.1.3   Geo-Indistinguishability Method

Step 1: Laplace Mechanism in Continuous Plane   Laplace noise is added to points to achieve geo-indistinguishability in the continuous plane. When user's actual location is $x_0 \in R^2$, planar Laplace noise will be added to the point. Equation (12.5) shows the mechanism:

$$K(x_0)(x) = \frac{\epsilon^2}{2\pi} e^{-\epsilon d(x_0, x)},$$ (12.5)

where $\frac{\epsilon^2}{2\pi}$ is a normalization factor.

Step 2: Discretizated Step to Achieve Geo-Indistinguishability in the Discrete Domain   One common way to analyze trajectories on a continuous domain (and to limit the size of the model) is using discretization on the space. After defining a grid $\mathscr{G}$ of discrete Cartesian coordinates, remap $\widehat{x} = K(x_0)(x)$ to the closest point x on $\mathscr{G}$.

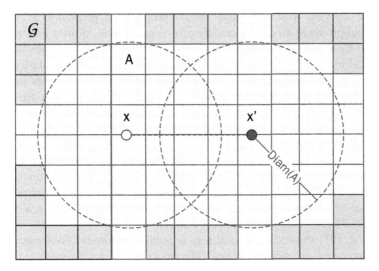

Fig. 12.3 Geo-indistinguishability mechanism

Step 3: Truncated Step to Obtain Finite Regions The Laplace mechanisms described in the previous sections have the potential to generate points everywhere in the plane, but in most cases, users are only interested in location with finite region. In this step, a finite set $A \in R^2$ is defined. All reported points $\hat{x}$ will be mapped to the closest point in $A \bigcap \mathcal{G}$.

Figure 12.3 shows the three steps mechanism. The orange region represent the planar Laplace noise added to points x and x'. The step 2 decretized the Laplace noise in grid $\mathcal{G}$, and dot line circles denote the truncated region A in step 3. Geo-indistinguishability is an specific instance of differential privacy, uses an arbitrary metric between secrets.

12.3.2   Trajectory Privacy

The trajectory has several properties that make the privacy preserving a challenge [97]. Firstly, individual's trajectories are highly unique. This indicates the difficulty of achieving properties such as k-anonymity, where at least k individuals have the same trajectory. Secondly, the trajectory is predictable in most cases. The release of an individual's location data can make him susceptible to privacy attacks. Existing techniques for privacy-preserving location data release are derived using partition-based privacy models, which have been shown failing to provide sufficient privacy protection. Differential privacy is considered as a suitable privacy model that can provide a rigorous privacy guarantee.

Ho et al. [99] took advantage of the property of prediction in trajectory release and proposed an ϵ-differentially private geographic location pattern mining approach by a partition-aggregate framework. The framework utilizes the spatial decomposition to limit the number of records within a localized spatial partition and then applies noise-based clustering to discover the interesting patterns in location datasets. Laplace noise is added in both steps with the aim to mask the count of records in a region and the centroid of these records. Chen et al. [40] presented a data-dependent solution for sanitizing large-scale trajectory data. They developed a constrained inference technique to increase the resulting utility. He et al. [97] presented a system to synthesize mobility data based on raw GPS trajectories. They discretize raw trajectories using hierarchical reference systems to capture individual movements at differing speeds. They then propose an adaptive mechanism to select a small set of reference systems to construct prefix tree counts. Lastly, a direction-weighted sampling is applied to improve utility.

He et al. [97] showed a practical way to publish synthesize GPS trajectories in the constraint of differential privacy. The major difference between independent location points and the trajectory is the correlation between locations in trajectory. The key idea of He et al.'s work is map the original locations to anchor points in various grid, defined as reference systems. And create sequential data is the order Markov process that generates the next location of trajectory based on the previous k locations. Laplace noise is added to the transitive probabilities to achieve differential privacy.

They consider D as a trajectory dataset, in which each individual has a personal trajectory table, denoted by PT with each tuple representing a regular trajectory, t. It is possible for an individual to have multiple trajectories, of varying lengths, where length is defined as the number of points in the trajectory. A uniform grid $\mathscr{G}_v$ is generated over the space and choosing the centroid as anchor points, where v denotes the length of the side of each grid cell. They use $c(t, x)$ to denoted occurrence of x in t. Figure 12.4 shows an example. In a grid G, suppose a user has two trajectories t_1 and t_2, both trajectory has passed the locations x_i, $c(t_1, x_i) = 1$, while $c(u, x_i) = 2$. In the figure, point a is an anchor point that may replace nearby points x_j. The background coordinate is another G with the same trajectories but different resolution.

Two major steps involved into the DPT method. Step 1: Decritization by hierarchical reference systems: In this step, Markov processes is first applied to model correlations between contiguous locations in a trajectory.

Definition 12.2 (Markov Process) A regular trajectory $< x_1, \ldots, x_d >\in \mathscr{X}$ is said to follow an order φ Markov process if for every $\varphi \leq i \leq d, x \in \mathscr{X}$

$$Pr(x_{i+1} = x | x1 \ldots xi) = Pr(x_{i+1} = x | x_{i-\varphi+1} \ldots x_i). \tag{12.6}$$

The probability $Pr(x_{i+1} = x | x_{i-\varphi+1} \ldots x_i)$ is a transition probability that is estimated by φ- and $\varphi + 1$ length counts.

Fig. 12.4 Trajectory count

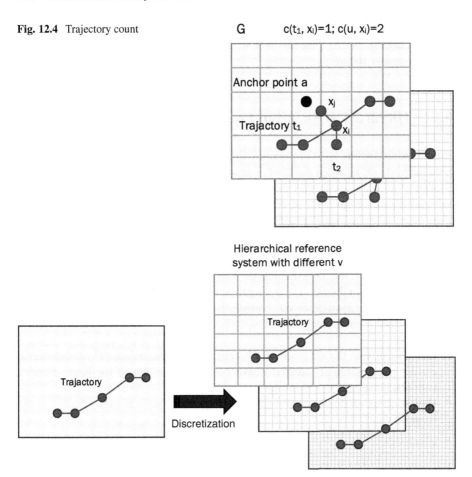

Fig. 12.5 Hierarchical reference system

The size of v in grid G is tricky as there are several problems: (1) a user may travel in different speeds so that some anchor points may not be estimated in a same scale. (2) Large v leads to huge utility loss, while small v leads to significant number of parameters in the Markov process. Therefore, a hierarchical reference system, shown as Fig. 12.5, is applied to variate the size of v, with geometrically increasing resolutions. Given a point x, there are two ways to map x to the next anchor point a_{i+1} where (1) a_{i+1} is a neighboring cell of a_i in the same G; or (2) a_{i+1} is the parent or a child of a_i in a different G. After mapping trajectory in grids, step 1 outputs mapped trajectory such as $t = a_1, \ldots a_d$, in which anchor points may be derived from different grid with various v.

Step 2 selects the private model. One particular trajectory may have various mapping in the first step. This step will select the most suitable perturbed trajectory based on the transitive probabilities in Markov process. To preserve privacy, Laplace noise is added to the count of locations, which changes the transitive probabilities.

As noise may hide the direction of the trajectory, weights are calculated by previous directions in a windows size w. Then the weighted sampling process will help to improve the utility.

12.4 Hierarchical Snapshot Location Publishing

When addressing the issues of privacy in location dataset, the main challenge is to obtain a trade-off between the level of privacy and the utility of the released dataset. Most traditional privacy preserving methods lack of a rigid privacy guarantee on the released dataset. Hence, Xiong et al. [242] proposed a *Private Location Publishing* algorithm, with the aim to preserve the privacy for individuals while maximizing the utility of the released dataset. Specifically, they attempt to address the following issues:

- How to decrease the large magnitude of noise when using differential privacy? In a sparse location dataset, the key method to decreasing the noise is to shrink the scale of the randomization mechanism. Previous work focuses on methods that generalized a particular location to a vague one, but this results in high utility loss.
- How to calibrate the sensitivity for hierarchical structure dataset? As mentioned earlier, the traditional sensitivity will be very large due to the large distance between locations.
- How to design a release mechanism that retains the semantic information of the location data?

12.4.1 Hierarchical Sensitivity

Before proposing the *hierarchical sensitivity*, they first analyze the source of redundant noise according to the definition of *sensitivity* in differential privacy. For a location dataset, the distance between locations is continuous, in order to preserve privacy, the sensitivity of a query has to be calibrated by the maximal distance between locations, which could completely mask the true distance and render the utility of dataset useless. They observe that a location dataset maintains a hierarchical structure, which defines the different semantic on each level. For example, country, city and street can be considered as levels of the location dataset, and each level has its own semantic. In this hierarchical structure, users may have diverse requirements on different levels of location. Some of them just need to hide the street, or hide the city, few of them need to hide the county. Hence, they can associate the hierarchical structure of the dataset with users' various requirements.

Based on this observation, they define the *hierarchical sensitivity* for differential privacy mechanism, it is calibrated by the maximal distance between two locations on the same level of location. Let L represent the level of the structure that user intends to preserve, they then have the follow definition:

Definition 12.3 (Hierarchical Sensitivity)   For a given L, the hierarchical sensitivity of L is

$$HS_L = \max_L d(t_i, t_j), \qquad\qquad (12.7)$$

where $d(t_i, t_j)$ represent the distance between t_i and t_j.

The user can choose their requirement of the privacy level L and the *hierarchical sensitivity* is generated according to Definition 12.3. For example, to preserve the `city` level privacy, the sensitivity is measured by the maximal distance within this city. Then the differential privacy mechanism will mask the `city` of the user rather than the `country` of the user. By this way, the randomized dataset can reduce the amplitude of the noise.

They illustrate the concept of hierarchical sensitivity using an example. Giving a map consisting of three difference cities, which denoted by three circles C_1, C_2 and C_3, as shown in Fig. 12.6. The distances between cities are $d(C_1, C_2)$, $d(C_2, C_3)$ and $d(C_1, C_3)$, respectively. Let nodes S_1, S_2 and S_3 in each circle represent streets, and $d(S_1, S_2)$, $d(S_2, S_3)$ and $d(S_1, S_3)$ be the distances between corresponding streets. If a user chooses a city level privacy, it means that he wants to hide the city he has travelled. The neighboring dataset should be a graph with revising the location of a city. The hierarchical sensitivity is then measured by the maximal distance between cities: $HS_{city} = \max(d(C_1, C_2), d(C_2, C_3), d(C_1, C_3))$. Similarity, if the use chooses the street level privacy, there will be $HS_{street} = \max(d(S_1, S_2), d(S_2, S_3), d(S_1, S_3))$.

Fig. 12.6 Distance between streets and cities

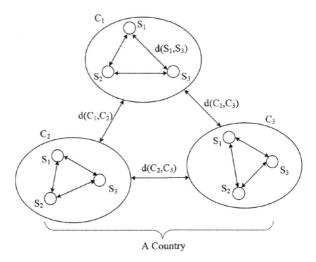

A Country

They consider the privacy level L as a configuration in a software or a service. When a user intends to use a software with location information or a location based service, he/she can set this parameter based on his/her own preference. It can either be configured before the service or during the service. They can also set a default value if the user skips this setting procedure. In the following sections, they assume that the privacy level L and the related hierarchical sensitivity have been generated in advance. They then simplify the notation from HS_L to HS.

12.4.2   Overview of Private Location Release

In a location dataset D, the information about a user, say u_a, can be represented by a profile $P(u_a) =< T(u_a), W(u_a) >$, where $T(u_a)$ is the set of all locations in this dataset, and $W(u_a)$ is the weight vector representing the frequency distribution of these locations. If an adversary has partial information on $P(u_a)$, he/she may re-identify user u_a in the location dataset by searching those known locations. More background knowledge will lead to higher probability of re-identifying a user. Hence, traditional privacy approaches hardly provide sufficient protection to users, due to the difficulty in modeling background knowledge [84].

Differential privacy assumes that all locations in the dataset have same probabilities to appear in a user's profile. Specifically, locations in $P(u_a)$ are represented by $T(u_a) = \{t_1, \ldots, t_{|T|}\}$, where $|T|$ is the total number of locations appeared in D, and the weights are denoted as $W(u_a) = \{w_1, \ldots, w_{|T|}\}$, where $w_i = 0$ indicates that location t_i has never been visited by this user. *Differential privacy* then utilizes the randomized mechanism to add noise into the weight $W(u_a)$ and releases a noisy profile $\widehat{P}(u_a) =< T(u_a), \widehat{W}(u_a) >$. In this case, $W(u_a)$ is a sparse vector as a user can only visit a limited number of locations. When applying the randomized mechanism, $\widehat{W}(u_a)$ will contain a large amount of noise because lots of weights in $W(u_a)$ will change from zero to a positive value.

One strategy to reduce the noise is to shrink the randomized domain, which refers to the diminished number of zero weights in the profile. To achieve this objective, they structure the locations on each level into η *clusters* and each user is represented by a *cluster-based* profile $P_C(u_a) =< T_C(u_a), W_C(u_a) >$, where the subscript C means clustering based. In the profile, $T_C(u_a) = \{T_{C_1}(u_a), \ldots, T_{C_\eta}(u_a)\}$ is the locations grouped by clusters, where $T_{C_i}(u_a)$ represents the ith location cluster of u_a. $W_C(u_a) = \{w_{C_1}(u_a), \ldots, w_{C_\eta}(u_a)\}$ is a weight vector representing the frequency summary grouped by the clusters. Namely, $w_{C_j}(u_a)$ is the total frequency of the jth location cluster visited by u_a. Then the noise is added to the frequency based on each cluster instead of every particular location. Compared to $W(u_a)$, $W_C(u_a)$ is much less sparse. Because the noise added to each $w_{C_j} \in W_C(u_a)$ equals to $w_j \in W(u_a)$, the total noise added to $W_C(u_a)$ will significantly decreased.

In this section, they propose a *Private Location Release (PriLocation)* algorithm to address the privacy issues in location dataset. They first present an overview of the algorithm, then provide details of its operations.

12.4.3   Private Location Release Algorithm

The *PriLocation* algorithm aims to publish all users' profiles by masking the exact locations and weights under the notion of differential privacy. Three private operations are introduced to ensure that each individual in the releasing dataset cannot be re-identified by an adversary.

- *Private Location Clustering*: This creates location clusters and masks the exact number of locations as well as the centers of each cluster. From the clustering output, the adversary cannot infer to which cluster a location exactly belongs.
- *Cluster Weight Perturbation*: This operation masks the weights of the locations in a user's profile to prevent an adversary from inferring how many locations a user has visited in a certain cluster.
- *Private Location Selection*: This aims to mask a user's profile that an adversary cannot infer the locations visited by this user.

Based on these private operations, the proposed *PriLocation* algorithm generates a sanitized location dataset, and its pseudocode is provided in Algorithm 1. Firstly, step 1 divides the privacy budget into three parts, corresponding to the three private operations. Step 2 groups all locations into η clusters, and in step 3 the weight for each cluster is perturbed by Laplace noise. After privately selecting the new locations to replace the original ones in step 4, the sanitized dataset $\widehat{D}$ is released in the last step.

Algorithm 1 Private Location Release (PriLocation) Algorithm

Require: D, privacy parameter ϵ, η.
Ensure: $\widehat{D}$
    1. Divided privacy budget into $\epsilon/2$, $\epsilon/4$ and $\epsilon/4$;
    2. *Private Location Clustering*: cluster locations into η groups with $\epsilon/2$ privacy budget;
    for each user u_a **do**
        3. *Cluster Weight Perturbation*: add Laplace noise to the group weights with $\epsilon/4$ privacy budget;

$$\widehat{W}(u_a) = W(u_a) + Laplace(\frac{4}{\epsilon})^{\eta}.$$

        for each Cluster C_η in $P(u_a)$ **do**
            4. *Private Location Selection*: Select locations according to the $\widehat{W}(u_a)$ with $\epsilon/4$ privacy budget;
        end for
    end for
    5. Output $\widehat{D}$;

These three private operations simultaneously guarantee a fixed ϵ-differential privacy and retain the acceptable utility of dataset. Details for the *Private Location Clustering* operation is presented in Sect. 12.4.3.1, followed by the *Cluster Weight Perturbation* in Sect. 12.4.3.2 and the *Private Location Selection* in Sect. 12.4.3.3.

12.4.3.1   Private Location Clustering

In this subsection, they describe the *Private Location Clustering* operation that privately groups locations into clusters. *Private Location clustering* categorizes unstructured locations on each level into groups to shrink the randomized domain for privacy purposes. According to the notion of differential privacy, this operation ensures that deleting a location will not significantly affect the clustering results. This means that an adversary cannot infer to which group a location belongs from the clustering output. This objective can be achieved by adding Laplace noise in the distance measurement during the clustering process.

As one of the most popular models, *k-means* is easy to be extended for privacy preserving, especially for differential privacy [28]. Therefore, they apply *k-means* as the baseline clustering algorithm and introduce the differential privacy to generate a private cluster algorithm. Blum's initial work in *SuLQ* framework claims that a private clustering method should mask the cluster center and the number of records in each cluster. Following this work, they conceptualize the *Private Location Clustering* in two steps:

- Defining a quantitative measure of distance between locations.
- Adding noise in each iteration to privately cluster all locations.

As the location is defined by the *longitude* and *latitude*, The distance between locations can be measured by *Euclidean* distance:

$$d(t_i, t_j) = \sqrt{(x_i - x_j)^2 + (y_i - y_j)^2}. \tag{12.8}$$

The next step introduces differential privacy into *k-means*, which is essentially an iterated algorithm for grouping data observations into η clusters. Let c_l denote the center of the cluster C_l. Equation (12.9) shows the objective function g, which measures the total distance between the location and the cluster center it belongs to.

$$g = \sum_{i=1}^{|T|} \sum_{l=1}^{\eta} \gamma_{il} d(t_i, c_l), \tag{12.9}$$

where γ is an indicator defined as follows.

$$\gamma_{il} = \begin{cases} 1 & t_i \in C_l \\ 0 & t_i \notin C_l \end{cases} \tag{12.10}$$

When combined with differential privacy, Laplace noise is calibrated by the *hierarchical sensitivity HS* of the objective function G and the privacy budget. For the privacy budget, they separate $\epsilon/2$ into p parts, where p is the number of iterations. So the private objective function $\widehat{G}$ is defined as following:

$$\widehat{G} = \sum_{i=1}^{m} \sum_{l=1}^{\eta} \gamma_{il} d(t_i, c_l) + Laplace\left(\frac{2p \cdot HS}{\epsilon}\right). \tag{12.11}$$

After p iterations, the *Private Location Clustering* outputs $\widehat{C} = \{C_1, \ldots, C_\eta\}$. Details of this operation are shown in Algorithm 2.

12.4.3.2 Cluster Weight Perturbation

After generating the cluster-based user profile $P_C(u_a)$, Laplace noise will be added to mask the counts of locations in each cluster.

$$\widehat{W}_C(u_a) = W_C(u_a) + Laplace\left(\frac{4}{\epsilon}\right)^\eta. \tag{12.12}$$

Noise added on the weight $W_C(u_a)$ implies the revision of the locations in $T_C(u_a)$. Positive noise indicates that new locations are added to the $T_C(u_a)$, while negative noise indicates that some locations have been deleted from the list. For positive noise in the cluster C_l, the operation will choose the location close to the cluster center. For negative noise, the operation will delete the location with the largest distance to the cluster center. Namely,

---

Algorithm 2 Private Location Clustering Operation

Require: Location set T, privacy parameter $\epsilon/2$, iteration round p, numbers of clusters η
Ensure: $\widehat{C} = \{C_1, \ldots, C_\eta\}$
   1. Randomly select centers $c_1, \ldots, c_\eta$;
  for 1:p **do**
     2. Assign all the locations to η clusters and get their indicators;
     3. Measure the private objective function

$$\widehat{G} = \sum_{i=1}^{m} \sum_{l=1}^{\eta} \gamma_{il} d(t_i, c_l) + Laplace(\frac{2p \cdot HS}{\epsilon});$$

     4. Update clustering centers according to $\widehat{G}$;
  end for
   5. Output $\widehat{C} = \{C_1, \ldots, C_\eta\}$;

$$\widetilde{T}_{C_l}(u_a) = T_{C_l}(u_a) + t_{new}, \tag{12.13}$$

where $t_{new} = \arg\min_{t_i \in C_l} d(t_i, c_l)$.

$$\widetilde{T}_{C_l}(u_a) = T_{C_l}(u_a) - t_{delete}, \tag{12.14}$$

where $t_{delete} = \arg\max_{t_i \in C_l} d(t_i, c_l)$.

After perturbation, they use $\widetilde{P}_C(u_a) =< \widetilde{T}_C(u_a), \widehat{W}_C(u_a) >$ to represent the noisy cluster-based user profile. However, the $\widetilde{P}_C(u_a)$ still has the high probability to be re-identified because it retains a major part of the original locations. The next operation will replace all locations in $\widetilde{T}_C(u_a)$ to preserve privacy.

12.4.3.3 Private Location Selection

The *Private Location Selection* operation replaces original locations with selected new locations. The challenge is *how to select a new location* from the related clusters. For a location $t_i \in \widetilde{T}_{C_l}(u_a)$, uniformly random location selection within $\widetilde{T}_{C_l}(u_a)$ is unacceptable due to significant utility detriment. The intuitive approach to retaining utility is to replace with the most similar location. However, this approach is also insecure because the adversary can easily figure out the location most similar using simple statistical analysis. Consequently, the *Private Location Selection* needs to: (1) retain the utility of locations, and (2) mask the similarities between locations.

To achieve these requirements, *Private Location Selection* adopts the exponential mechanism to privately select locations from a list of candidates. Specifically, for a particular location t_i, the operation first locates the cluster C_l to which it belongs, and all the locations in $\widetilde{T}_{C_l}(u_a)$ are then included in a candidate list I. Each location in I is associated with a probability based on the *score function* and its related *sensitivity*. The selection of locations is performed based on the allocated probabilities.

The *score function* is defined by the distance between locations. They define the score function q for a candidate location t_j with the target location t_i as follows:

$$q_i(I, t_j) = (HS - d(t_i, t_j)), \tag{12.15}$$

where I is location t_i's candidate list, and $t_j \in I$ is one candidate location for replacement. Each location t_j has a score according to Eq. (12.15).

The *sensitivity* for score function q is measured by the maximal change in the distance between t_i and t_j. Here, they will use the *hierarchical sensitivity HS*.

Based on the *score function* and the *sensitivity*, the probability arranged to each location t_j is computed by Eq. (12.16) with the privacy budget $\frac{\epsilon}{4}$.

$$Pr_{t_j \in I}(t_j) = \frac{\exp\left(\frac{\epsilon \cdot q_i(I, t_j)}{8 \cdot HS}\right)}{\sum_{t_j \in I} \exp\left(\frac{\epsilon \cdot q_i(I, t_j)}{8 \cdot HS}\right)}. \tag{12.16}$$

They then select a location t_j from C_l to replace the t_i according to this probability. Eventually, the algorithm output $\widehat{T}_C(u_a)$. The pseudocode of *Private Location Selection* is presented in Algorithm 3.

Algorithm 3 Private Location Selection

Require: $\frac{\epsilon}{4}, \widetilde{T}_C(u_a)$,

Ensure: $\widehat{T}_C(u_a)$

   for each location t_i in $\widetilde{T}_C(u_a)$ **do**
      1. locate the t_i in cluster C_l;
      for each location t_j in C_l **do**
         2. Allocate probability as:

$$\frac{\exp\left(\frac{\epsilon \cdot q_i(l,t_j)}{8 \cdot HS}\right)}{\sum_{j \in C_l} \exp\left(\frac{\epsilon \cdot q_i(l,t_j)}{8 \cdot HS}\right)}.$$

   end for
      3. Select a location t_j from C_l to replace the t_i according to the probability;
   end for
      4. Output $\widehat{T}_C(u_a)$;

12.4.4   Utility and Privacy

12.4.4.1   Utility Analysis

The utility of dataset $\widehat{D}$ highly depends on the user profile. The closeness between the user's original profile and the perturbed profile is the key factor that determines the utility level. Given a target user u_a, they set the original profile P_{u_a} as a baseline. By comparing the replaced locations in the user's profile $\widehat{P}_{u_a}$ with the corresponding ones in the baseline P_{u_a}, they can evaluate the utility level of the proposed algorithms. The distance between P_{u_a} and $\widehat{P}_{u_a}$ is referred to as the *distance error*, which is a direct measurement on the difference between the locations before and after the randomization.

$$DE_{u_a} = \frac{\sum_{t \in \widehat{T}_C(u_a)} d(t, \widehat{t})}{HS \cdot |\widehat{T}_C(u_a)|}, \tag{12.17}$$

where $\widehat{t}$ is the new location replacing the original t and HS is a fixed pre-determined *hierarchical sensitivity*.

For the entire dataset $\widehat{D}$, they have the average *distance error* as Eq. (12.18).

$$DE = \frac{1}{|U|} \sum_{u_a \in U} (DE_{u_a}). \tag{12.18}$$

The error measurement can be applied to evaluate the utility of $\widehat{D}$ in terms of *distance error*. They consider the *PriLocation* algorithm as query set F and prove that with a high probability it is less than a certain value α. The following Theorem 5 provides the bound of α, which implies the minimal distance error of the *PriLocation* algorithm. The α indicates the least utility loss that needs to be sacrificed, if a certain level of privacy needs to be achieved.

Theorem 5 *For any user $u_a \in U$, for all $\delta > 0$, with probability at least $1 - \beta$, the distance error of the released dataset is less than α. The lower bound of α is presented by Eq. (12.19)*

$$\alpha \leq \max_{u_a \in U} \frac{\sum_{t_i \in \widehat{T}_C(u_a), t_j \in C_{t_i}} E(d(t_i, t_j))}{HS \cdot |\widehat{T}_C(u_a)| \cdot \beta}, \tag{12.19}$$

where $E(d(t_i, t_j))$ denote as the expectation of $d(t_i, t_j)$.

Proof The *distance error DE* is proportion to the distance between the original location and the selected location. Given a user u_a who has a set of locations $T(u_a)$, for each original location t_i, the probability of being replaced by a privately selected location t_j is defined by Eq. (12.16). Based on this probability, they can estimate the scale of DE_{u_a} for user u_a.

According to *Marlkov's inequality*, for user u_a, they have

$$Pr(DE_{u_a} > \alpha_a) \leq \frac{E(DE_{u_a})}{\alpha_a}$$

$$\Rightarrow Pr(DE_{u_a} \leq \alpha_a) > 1 - \frac{E(DE_{u_a})}{\alpha_a}.$$

According to Definition 14.9, they estimate the $Pr(DE_{u_a} \leq \alpha_a)$. Let $1 - \frac{E(DE_{u_a})}{\alpha_a} = 1 - \beta$, they have

$$\alpha \leq \max_{u_a \in U} \frac{\sum_{t_i \in \widehat{T}_C(u_a), t_j \in C_{t_i}} E(d(t_i, t_j))}{HS \cdot |\widehat{T}_C(u_a)| \cdot \beta}. \tag{12.20}$$

For all users, α is determined by the maximal value, $\alpha = \max_{u_a \in U} \alpha_a$.

They can have one step to further estimate the expectation $E(DE_{u_a})$. According to Eq. (12.16), there will be

$$E(DE_{u_a}) = \sum_{t_i \in \widehat{T}_C(u_a), t_j \in C_{t_i}} \frac{d(t_i, t_j)}{HS \cdot |\widehat{T}_C(u_a)|} \cdot \frac{\exp(\frac{\epsilon \cdot (HS - d(t_i, t_j))}{8|\widehat{T}_C(u_a)|HS})}{\rho_i},$$

where ρ is the normalization factor depending on the cluster to which t_i belongs to.

The proof shows that the *distance error* for each user mainly depends on the privacy budget and the normalization factor ρ_i. According to Eq. (12.16), the normalization factor ρ_i for a particular location t_i is defined as

$$\rho_i = \sum_{t_i \in \widehat{T}_C(u_a), t_j \in C_{t_i}} \exp\left(\frac{\epsilon \cdot (HS - d(t_i, t_j))}{8|\widehat{T}_C(u_a)| \cdot HS}\right). \tag{12.21}$$

Table 12.3 Privacy budget allocation in *PriLocation* algorithm

Operations	Privacy budget
Private location clustering	$\epsilon/2$
Cluster weight perturbation	$\epsilon/4$
Private location selection	$\epsilon/4$

Therefore, the size of ρ_i is depended on the cohesion inside cluster C_{t_i}, in which the compact cohesion results in small ρ_i and less *distance error*. Further analysis shows that the cohesion is determined by the privacy budget ϵ in the *private location clustering* operation. It can be concluded that *the privacy budget affects on the utility level of* PriLocation.

12.4.4.2   Privacy Analysis

The *PriLocation* algorithm contains three private operations: *Private Location Clustering*, *Cluster Weight Perturbation* and *Private Location Selection*. The privacy budget ϵ is consequently divided into three pieces, as illustrated in Table 12.3.

Because the *Private Location Clustering* operation is performed on the entire dataset and will have effect on all users, they allocate more privacy budget ($\epsilon/2$) than other two operations. The other two operations only perform on individuals, and less privacy budget are required. They allocate the rest $\epsilon/2$ to these two operations ($\epsilon/4$ for each).

Based on the privacy compositions and the privacy budget allocation in Table 12.3, they measure the privacy level of our algorithm as follows:

- The *Private Location Clustering* operation is performed on the whole dataset with the privacy budget $\frac{\epsilon}{2}$. According to the parallel composition, this operation preserves $\frac{\epsilon}{2}$-differential privacy.
- The *Cluster Weight Perturbation* applies the Laplace mechanism to the weights of clusters. The noise is calibrated by $Lap\left(\frac{4}{|T(u)|\cdot\epsilon}\right)^{\eta}$ and preserves $\frac{\epsilon}{4}$-differential privacy for each user. Furthermore, as a user's profile is independent, replacing a user's locations has no effect on other user profiles. The *Cluster Weight Perturbation* preserves $\frac{\epsilon}{4}$-differential privacy as a whole.
- The *Private Location Selection* adopts the Exponential mechanism. For one user u, each location in the profile is replaced by a privately selected location until all locations have been replaced. Each selection is performed on the individual location, therefore according to parallel composition, for each user, the selection guarantees $\frac{\epsilon}{4}$-differential privacy. Similar to the previous operation, every user profile can be considered as a subset of the entire location dataset. Thus, the *Private Location Selection* guarantees $\frac{\epsilon}{4}$-differential privacy.

Consequently, they can conclude that the proposed *PriLocation* algorithm preserves ϵ-differential privacy.

12.4.5 Experimental Evaluation

12.4.5.1 Datasets

To obtain a thorough comparison, they conduct the experiment on four datasets: GeoLife, Flickr, Diversification and Instagram. All datasets are structured in the form of (User, Country, City, Street, Latitude and Longitude).

GeoLife   *GeoLife* is a location-based social networking service, which enables users to share life experiences and build connections among each other using human location history. GeoLife dataset contains 17,621 traces from 182 users, moving mainly in the north-west of *Beijing, China,* in a period of over 5 years (from April 2007 to August 2012).

Flickr   The *Flickr* dataset is crawled from www.flickr.com. The dataset contains 1692 individual users, with 26,616 records. The City attribute of the dataset covers *New York, Paris, Melbourne, Hong Kong,* and *Macau.*

Div400   This dataset was validated during the *2013 Retrieving Diverse Social Images Task* at the *MediaEval Benchmarking Initiative for Multimedia Evaluation* footnotehttp://www.multimediaeval.org/. It contains 43,418 records related to 396 locations.

Instagram   They also crawled a sample of public *Instagram* photos with locations.[1] The dataset contains 2015 individual users, with 28,767 records.

12.4.5.2 Estimation of Distance Error

To maintain the consistency with previous research, they thoroughly compare the *distance error* of *PriLocation* with the traditional differential privacy (DP) algorithm as well as the *k-anonymity* approach on the four datasets. For the *PriLocation* algorithm, they first fix the *street* privacy level, and set ϵ from 0.1 to 1.0 with a step of 0.1. They set the number of cluster $\eta = 10, 40$ and 80. For the traditional differential privacy mechanism, they also set ϵ from 0.1 to 1.0 with a step of 0.1. For the *k-anonymity* approach, as it adopt k to control the privacy level, rather than ϵ, they set two empirical values, $k = 10$ and 50.

Figure 12.7 shows the results on those four datasets. It can be observed that the *distance error* of the *PriLocation* algorithm in a variety of η is less than that of the traditional differential privacy with different privacy budgets, and this indicates that *PriLocation* outperforms naive differential privacy on all the datasets. Specifically, the *PriLocation* algorithm obtains a considerably lower *distance error* when $\epsilon = 1$. For example, in Fig. 12.7a, when $\eta = 80$ and $\epsilon = 1$, the *distance error* is 0.0239, which is 69.14% lower than that of traditional differential privacy. Even in a higher

---

[1]http://instagram.com.

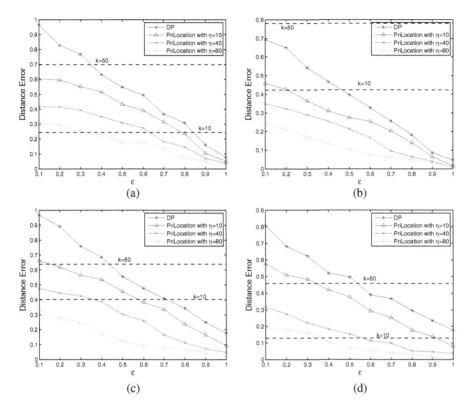

Fig. 12.7 *Distance Error* on different datasets with `Street` level privacy. (**a**) *Distance Error* in *Flickr*. (**b**) *Distance Error* in *GeoLife*. (**c**) *Distance Error* in *Div400*. (**d**) *Distance Error* in *Instagram*

privacy level ($\epsilon = 0.1$), the *distance error* is still lower than that of the traditional differential privacy by 67.81% when $\eta = 80$. This trend retains when η equals to other values, thus illustrates the effectiveness of *PriLocation* in terms of the distance information retained in the randomized dataset. Similar trends can also be observed in Fig. 12.7b–d. All figures show that *PriLocation* obtains a stable reduced *distance error* comparing with the naive differential privacy.

They also compare the *PriLocation* algorithm with *k-anonymity* approach. Figure 12.7 shows that *k-anonymity* has diverse performances in different datasets. For *Flickr* dataset, Fig. 12.7a shows that the *distance error* of *k-anonymity* is 0.2446 when $k = 10$. This result outperforms *PriLocation* with $\epsilon < 0.4$. However, when $\epsilon > 0.4$, $\eta = 80$, *PriLocation* outperforms *k-anonymity* with $k = 10$. When $k = 50$, *k-anonymity* has higher *distance error* ($DE = 0.6980$) than all settings of *PriLocation*. In Fig. 12.7b. when $k = 10$, *k-anonymity* still has higher *distance error* comparing with *PriLocation*. When $k = 50$, *k-anonymity* has a poor performance ($DE = 0.7802$) comparing with *PriLocation*. They can observe that the performance of *k-anonymity* depends on the size of the datasets.

A larger dataset with smaller value of k has a better performance than a smaller dataset with larger k. *Flickr* dataset has 1692 individual users and *GeoLife* only contains 182 users, so that the performance of *k-anonymity* in *Flickr* is better than in *GeoLife* with the same value of k. This observation is confirmed by Fig. 12.7c, d. They shows that *Instagram* has a relative lower *distance error* comparing with *PriLocation* while *Div400* has higher *distance error*. This is because that *Instagram* has 2015 individuals while *Div400* only has 396. These results also demonstrate that *PriLocation* has a stable performance that is independents of the size of a dataset.

12.5 Summary

In the last a few years, providers of location-based services have collected a wealth of data of individuals and populations on the movement because of the popularity of such LBSs. However, the places that people visit will disclose extremely sensitive information such as their daily behaviours, home and work locations, preferences and habits, etc. Location privacy is, therefore, an urgent issue that needs to be addressed. The main aim of a location privacy LBS is to preserve as much as the quality of the desired services, while hindering the undesired tracking capacities of those services.

Although differential privacy has been considered as a promising solution to provide a rigid and provable privacy guarantee for data publishing, the straightforward differential privacy solution for location data still suffers from three important deficiencies: (1) it introduces a large volume of noise due to the sparsity of location dataset, which significantly affects the performance; (2) the definition of *sensitivity* in differential privacy can not be used directly in location data release because that it causes unacceptable noise.

This chapter describes three methods that apply differential privacy to achieve the aim of a location privacy LBS through various angles. Finally, a private publishing algorithm is proposed to randomize location dataset in differential privacy, with the goal of preserving users' identities and sensitive information. The algorithm aims to mask the exact locations of each user as well as the frequency that the user visit the locations with a given privacy budget.

Chapter 13
Differentially Private Spatial Crowdsourcing

13.1 Introduction

Crowdsourcing has been a successful business model ever since it was first introduced in 2006 [179]. It refers to employers outsourcing tasks to a group of people [179], and with the increasing use of smart equipment with multi-modal sensors. Spatial crowdsourcing (SC) [119] is a particular form of crowdsourcing where workers perform their assigned tasks at an exact location. The tasks might include taking photos, recording videos, reporting temperatures, etc. [216].

Within crowdsourcing platforms, tasks are uploaded by a requester to a spatial crowdsourcing server (SC-server) which assigns the tasks to registered workers using task-assigning algorithms. Workers submit their location in longitude/latitude form, to the SC-server, and taking the probability of a worker accepting a task into consideration, the server assigns each task to the most suitable workers to ensure a high assignment success rate with minimal cost. However, this process poses a privacy threat as the worker's location data may be released by the SC-server for further analysis, thus compromising the workers' privacy. Studies have demonstrated that published location information may be exploited to deduce sensitive personal information such as home location, political views, and religious inclinations [12, 100, 154].

Differential privacy has attracted attention in recent years in spatial crowdsourcing. Some algorithms, such as *DPCube* [238] and *AG* [145], have been proposed that allow the release of spatial datasets, while preserving privacy. The central idea of these algorithms is to decompose a large map into many cells of varying sizes and add noise into the count of users in each cell. These algorithms work well in scenarios of that answer range queries on location datasets, however, many weaknesses remain when applied directly in SC.

First, existing methods hide a worker's exact location in a cloaked region and suppose that workers are distributed uniformly within it. We argue that this assumption is invalid unless the size of the region is very small. In large regions,

© Springer International Publishing AG 2017   173
T. Zhu et al., *Differential Privacy and Applications*,
Advances in Information Security 69, DOI 10.1007/978-3-319-62004-6_13

Table 13.1 Application settings

Application	Crowdsourcing
Input data	Trajectory
Output data	Synthetic trajectory
Publishing setting	Non-interactive
Challenges	High sensitivity; data sparsity
Solutions	Adjust sensitivity measurement, using clustering to shrink the domain
Selected mechanism	Dataset partitioning
Utility measurement	Error measurement
Utility analysis	Union bound
Privacy analysis	Parallel composition
Experimental evaluation	performance measured by distance error

people may gather into small areas, because aggregation is a basic feature of human society. The algorithms also tend to decompose the whole map into a group of cells of varying sizes, yet maintain a similar worker count for each cell. Given a task in a large cell, an uneven distribution will cause a huge error when the density of workers around the task point is evaluated.

Second, existing methods generally consist of two steps—decomposing the map into a grid and adding noise for each cell—and each step consumes a certain amount of the privacy budget. Since the budget used for the noise-adding step is only a portion of the total budget, the noise volume will be significant.

To fix these weaknesses, this chapter first proposed a basic method of differentially private crowdsourcing. Based on that, the chapter present a reward-based spatial crowdsourcing, in which workers accept tasks in exchange for rewards and, at the same time, differential privacy is preserved. Table 13.1 shows the application settings.

13.2 Basic Method

13.2.1 Background of Crowdsourcing

SC can be categorized into two modes, worker selected tasks mode (WST) and server assigned tasks mode (SAT) [119].

In WST, the requester uploads tasks to the SC-server, and workers select their favorite tasks autonomously. WST is easy to implement but inefficient. Given workers always select the nearest tasks, the tasks with few workers nearby are accepted with low probability, which leads a low global task acceptance rate.

SAT is much more efficient. To ensure a high assignment success rate, workers must report their locations to the SC-server and tasks are assigned to the most suitable workers according to their position. Obviously, due to location exposure, this mode contains an inherent a privacy concern, and forms the main concern on the community.

The privacy issues in location-based services have been widely investigated. Many location-based privacy-preserving methods have been presented in recent decades, such as spatial cloaking [46, 163, 175] and dummy-based methods [124, 149]. Additionally, some studies have attempted to use these methods to solve location privacy problems in SC. For example, using the spatial cloaking method, Pournajaf et al. [183] presented a twofold approach. In the first step, cloaked areas are reported to SC-server rather than exact locations, and global task assignments are solved with uncertain locations. In the second step, workers refine the results of the first stage according to their exact location. Agir et al. [9] presented a location-obfuscation solution to preserve privacy. Obfuscation areas are generated dynamically, to replace exact locations, according to the parameters of personalized privacy requirements.

However, the main shortcoming of these methods is that their reliability is highly dependent on an adversary's background knowledge. A priori knowledge about an attack's objectives can be exploited to break privacy definitions [57].

13.2.2 Differentially Private Crowdsourcing Methods

To et al. [216] proposed a framework for protecting workers' locations by introducing the cellular service provider (CSP) as a trusted third party. As shown in Fig. 13.1, the CSP locates workers first and, because the SC server is assumed to be unreliable, the CSP's partition algorithm generates a private spatial decomposition (PSD) instead of publishing the workers' location information. The entire spatial region is decomposed into grid of indexed cells, each containing a count of the SC workers within. Laplace noise is then added to the count of each cell according to the definition of DP. Thus, a PSD represents a sanitized distribution of workers in the SC region. Anyone, including the SC-server, is unable to identify a user's

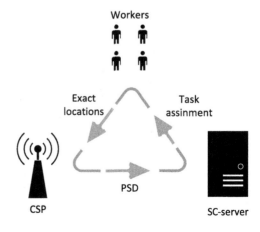

Fig. 13.1 An framework for private spatial crowdsourcing

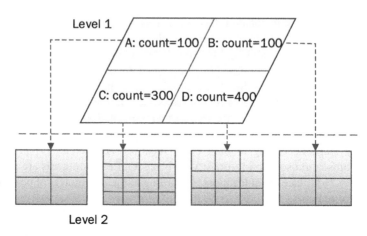

Fig. 13.2 Two-level adaptive grid method

location using a PSD. The PSD is then published to the SC-server instead of exact locations. Finally, the SC-server assigns the tasks to the workers according to the published PSD.

The entire process can be divided into two steps: Step 1: building the work private spatial decomposition (PSD) A PSD should be designed carefully for task assignment at the SC-server. They used a two level adaptive grid and variable cell granularity to avoid over or under-partition the space. The space is partitioned in larger granularity in the first level, and each cell is further partitioned based on the density of the workers. Figure 13.2 shows an example of the adaptive grid method. The first level uniformly partitions the region, while second level further partitions the region into smaller region according to the number of workers in each region. At this stage, Laplace noise is added to the count of workers at each level.

Step 2: task assignment. When a request for a task is posted, the SC-server queries the PSD and determines a geocast region, where the task is disseminated. The goal is to obtain a high success rate for task assignment. while at the same time preserve the system utility. To achieve the goal, three utility measurements have been proposed:

- Assignment Success Rate (ASR). ASR measures the ratio of tasks accepted by a worker to the total number of task requests.
- Worker Travel Distance (WTD). The total distances of workers need to travel to complete the task.
- Average number of notified workers (ANW). A significant metric to measure overhead of the system.

Based on the guide of these utility measurements, the geocast region should (1) contain sufficient workers such that task is accepted with high probability, and (2) keep a small size. The geocast region construction algorithm initially select a cell that covers the task, and determines its utility values. If the utility values are lower than thresholds, the algorithm expands the region by adding neighboring cells, until either when the utility of the obtained region exceeds the utility threshold. The construction algorithm is a greedy heuristic, as it always chooses the candidate cell that produces the highest utility increase at each step.

To et al. [217] presented a tool box, PrivGeoCrowd, to display the framework in a visual, interactive environment. System designers can use the tool box to obtain satisfactory results by tuning the parameters and allocation strategy.

13.3 Differential Privacy in Reward-Based Crowdsourcing

Though spatial crowdsourcing is a hot topic, only a limited number of studies have focused on its rewards. Dang et al. [52] considered the types of tasks given the varied expertise of workers. They defined a maximum-task minimum cost assignment problem and, by solving this problem, tasks could be assigned to the most expert workers, while minimizing the given reward. Wu et al. [233] concentrated on the relationship between crowdsourcing quality and payment. Their analyses are based on a 2D interior design task, and the evaluation results show that increasing monetary rewards does not significantly improve the average design quality, rather it increases the probability of excellent solutions from the crowd. Luo et al. [150] designed an incentive mechanism based on all-pay auctions. They transform the task assignment problem into a profit maximization problem, and introduce a contribution-dependent prize to model workers' contributions. Similarly, Shah-Mansouri et al. [199] propose an auction-based incentive mechanism, ProMoT, to maximize the profit of the platform while providing satisfying rewards to smartphone users. Thepvilojanapong et al. [215] took advantage of microeconomics, proposing SenseUtil, a participation-aware incentive mechanism with various utility functions to increase the number of sensors while keeping payments low.

While all the aforementioned works make contributions to their specified scenarios, few consider privacy issues when assigning a task. Moreover, most of these works regard reward allocation as an optimization problem, and do not integrate task assignment well.

Xiong et al. [241] presented a novel method for reward-based SC with a differential privacy guarantee, that uses a contour plot as a new sanitized representation of location distribution. They also propose a reward-based allocation strategy for achieving a specified assignment success probability in reward-based SC applications.

The major idea of this method is to split the entire area into smaller regions and apply a differential privacy mechanism to each region. As mentioned in Sect. 13.1, splitting regions is a non-trivial task. Previous literature assumes that worker's locations are distributed uniformly, but this may not be the case in real-world scenarios. The proposed method adopts a contour plot to illustrate workers' locations and determine an assignment task radius. In this way, the method arranges tasks with a better success rate and a privacy guarantee.

They follow the framework that To et al. [217] provided, which has been shown in Fig. 13.1. The CSP collects the true locations of workers, and generates a contour plot with a DP guarantee. The plot is then submitted to the SC-server. Because the exact worker locations will not be submitted, the privacy of each worker is preserved. The SC-server estimates the location region for each task, according to the contour plot (a circle is used to represent this region), and all workers within the circle are selected for the task.

13.3.1 Problem Statement

The fundamental goal of crowdsourcing is to arrange tasks for a group of workers efficiently. Efficiency is normally measured by an assignment success rate, indicating that the method has to ensure that each task has a high probability of being accepted. The *assignment success probability* (*ASP*) and an *acceptance probability* are defined to illustrate this.

Definition 13.1 (Assignment Success Probability) Given a task, t, and a set of workers, P, ASP is the probability that at least one worker in P accepts task t.

In addition, the *expected assignment success probability*, E_{ASP}, is defined as the minimum threshold every task must satisfy.

Definition 13.2 (Acceptance Probability θ) $\theta = Pr(p$ accepts $t)$ shows the probability of a worker, p, accepting a task, t.

The acceptance probability is dominated by the reward, w, and the distance between p and t. For a fixed distance, a higher reward means a higher θ. Conversely, a shorter distance results in a higher θ when the reward has been determined. Because the SC-server is unawares of the true location of a worker, it allocates the task to several workers who are located in a particular region. The center of the region is the task's location, and any workers within this circle will be chosen for the task.

Therefore, giving an entire reward, W, how to generate the differentially private region with radius r for t while ensuring the ASP is larger than threshold E_{ASP}, is the problem that must be solved.

Notations are showed in Table 13.2.

13.3.2 Building a Contour Plot with DP Guarantee

To perform task assignment and guarantee a high ASP, the SC-server must be aware of workers' location distribution, but submitting accurate worker locations may violate personal privacy. Traditional privacy-preserving algorithms assume that workers are uniformly distributed in an area, yet in real-world, this may not be the case.

Table 13.2 Notations

t	A task that need to be assigned
T	A set of tasks
p	A worker registered for crowdsourcing
P	A set of workers
d	Euclidean distance between p and t
w	A piece of reward
W	Total amount of reward
θ	The probability that a worker accept a task
$\odot_r^t$	A circle region that has center t and radius r

The proposed method uses a contour plot to represent a workers' location distribution. The objective of applying a contour plot is to estimate the user density of a given point more accurately. In traditional grid-based approach, with the assumption of uniform distribution, the user density of a point is set to be the value of the cell which the point belongs to. Comparing to the grid-based approach, a contour plot also shows the trend of gradient variation of the user density, which can be used for calculating the user density of any point with a higher precision.

In practice, the map is decomposed into a $m \times n$ grid with equal-sized cells. The size of a cell can be set according to the scale of the map and the exact location distribution of workers, for example, a $0.3\,\text{km} \times 0.3\,\text{km}$ with the area $0.09\,\text{km}^2$. To add noise, the exact number of workers n_{exact} is counted in each unit cell, then Laplace noise is added as shown in Eq. (13.1)

$$n_{noisy} = n_{exact} + Laplace\left(\frac{S}{\epsilon}\right),\qquad(13.1)$$

where S is the sensitivity. After this step, a noisy count n_{noisy} is obtained for each unit cell. Finally, a contour plot can easily be created by connecting the cells with the same noisy count.

Figure 13.3 illustrates an example of building a contour plot with 6×6 unit cells. The entire area is split into 6×6 disjoint cells, shown with dash lines. The exact number of workers in each cell is labeled at the bottom right. Noise is added to each count number during the privacy-preserving process, resulting in noisy counts, as shown in Fig. 13.3b. Finally, the cells with the same noisy count are connected, e.g., the cells with noisy counts 36, 58, 65 in Fig. 13.3c. In this way, the contour plot in Fig. 13.3d is generated. This contour plot helps the server to allocate tasks with a differential privacy guarantee.

As an example, a worker distribution and count method is shown in Fig. 13.4. The horizontal and vertical directions represent longitude and latitude. The solid curves are contour lines with positive noisy counts, and the exact noisy counts are labeled on these contour lines.

It can be proven that the proposed contour method provides a solid DP guarantee. As shown in Fig. 13.3, the entire data domain is split into disjoint cells. Laplace

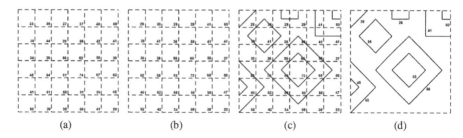

Fig. 13.3 Generation of a contour plot with 6 × 6 cells. (**a**) Exact Count. (**b**) Noisy Count. (**c**) Contour Lines. (**d**) Contour Plot

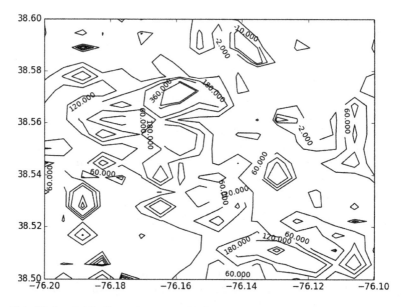

Fig. 13.4 Worker distribution and count mechanism

noise is added to the count of workers in each $cell_i$ independently, with a privacy budget of $\epsilon_i = \epsilon$. According to parallel composition (Definition 2.1), the sequence of the Laplace mechanism over the set of disjoint cells provides $\max\{\epsilon_i\}$-*differential privacy*. As each ϵ_i is equal to ϵ, the contour method is therefore ϵ-differentially private.

The contour plot is constructed at the CSP and submitted to the untrusted SC server, serving as an approximate distribution of workers. It prevent the SC server from identifying any worker with his/her location, while the SC server can assign the tasks to the workers according to the contour plot.

13.3.3   Task Assignment

After splitting the entire area into small cells, a radius, r, and a reward, w, are specified for each task. These two factors have a great impact on both the acceptance probability and the *ASP*.

13.3.3.1   Modeling Acceptance Probability and *ASP*

Intuitively, whether a worker will accept a task mainly depends on the reward, w, and the distance, d. Therefore, the acceptance probability θ ($0 \leq \theta \leq 1$) can be calculated by a function, f, with d and w, which is shown in Eq. (13.2)

$$\theta = f(w, d). \tag{13.2}$$

However, because of the complicated individual differences between workers, it's infeasible to define an optimal function to model the acceptance probability. Therefore, for simplicity, the features of the acceptance probability is first analyzed, and then a function that meet the requirements of these features is presented as a model of the acceptance probability.

Considering the natural properties of w and d, the function, f, has following features:

- $\theta \in [0, 1]$;
- If $w < w_0$ or d is larger than the *maximum travel distance* (*MTD*), which is the distance that a high percentage of workers are willing to travel [216]), then $\theta = 0$;
- When w ($w \geq w_0$) is fixed, θ will be increased with the decreasing of d;
- When d is fixed, θ will be increased when w is enhanced;
- When d is fluctuant, revising w ensures θ keeping on the same level.

The first statement clarifies that the non-negative probability, θ, is always less than 1. The second statement emphasises that w has a lowerbound of w_0, and d's upperbound is the *MTD*, which defines the necessary conditions for a user to accept a task. Intuitively, when the reward for a task is too low, or the distance is too long, the task has a very small probability of being accepted. A lower bound of reward, w_0, is defined as a threshold that a task will be accepted with a probability greater than 0 when the reward for the task is larger than w_0, otherwise the probability is equal to 0. Similarly, the *MTD* is the upper bound of the travel distance—a task will definitely be refused by all users when the distance of the task is larger than the *MTD*. The third and fourth statements indicate the relationship between acceptance probability and distance, and reward, respectively. Finally, the last statement specifies the tradeoff between the distance and the reward.

To model θ with the requirements mentioned above, we adopt the hyperbolic tangent function [11], defined in Eq. (13.3).

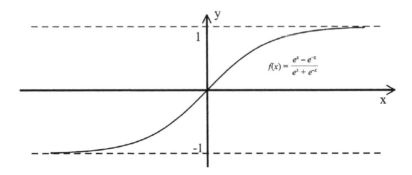

Fig. 13.5 Hyperbolic tangent function

$$y = \text{th}x = \frac{e^x - e^{-x}}{e^x + e^{-x}}. \tag{13.3}$$

It has the natural property that $y \in [0, 1)$ when $x \in [0, +\infty)$, and the horizontal asymptote, $y = 1$, when $x \to +\infty$. This means the function can be regarded as a mapping from non-negative numbers to probabilities. Hyperbolic tangent functions have been widely investigated because of these properties. In the field of engineering, Wang et al. [224] proposed a nonlinear transformation strategy based on hyperbolic tangent functions. Given a specific hyperbolic tangent function, the curve can be determined and the peak-to-average power ratio of transmitted signals in orthogonal frequency division multiplexing systems can be reduced effectively, with low degradation in the bit error rate. In addition, through translation and stretching, hyperbolic tangent functions can be transformed into the famous *logistic function* ($\text{th}x = 2logistic(x) - 1$), which has been widely studied and applied in the field of statistic [218]. The curve of the function is depicted in Fig. 13.5.

To model workers' acceptance probability, the hyperbolic tangent function can be applied as a framework of f. However, it will converge to 1 rapidly when x increases, i.e, the function value grows to its supremum in a small interval from the origin. the functions' values can be considered as probabilities. In practice, fast convergence will make the probabilities approximate to 1 for many x, and this does not fit with real-world scenarios. To overcome this shortcoming, we introduce parameter c_1 to control the scalability of the function, i.e.,

$$f = \text{th}(c_1 x). \tag{13.4}$$

The function f is considered as a bivariate function of distance, d, and reward, w, to characterize this combined relationship. Inspired by Ted's definition of *benefit-cost ratio* [213], widely used in investment analysis, the ratio of w and d of a task is a dominant factor that impacts the willingness of a user to accepting a task. Benefit-cost ratio is expressed as a ratio, where the numerator is the amount of monetary gain realized by performing a project, and the denominator is the amount it costs to execute the project. The higher the benefit-cost ratio, the better the project. In spatial crowdsourcing, μ is proportional to the ratio of w and d. It measures how attractive a task is for users,

$$\mu = c_2 \cdot \frac{w}{d}, \tag{13.5}$$

where c_2 is used for tuning the scale of the ratio. A larger value of μ implies a more attractive task, which may accepted with a higher probability.

According to Eq. (13.4), when $w \geq w_0$, we have

$$y = \text{th}(c_1 \cdot \mu) = \text{th}\left(c_1 \cdot c_2 \cdot \frac{w}{d}\right) = \text{th}\left(c \cdot \frac{w}{d}\right), \tag{13.6}$$

where c is the dot product of c_1 and c_2, namely, $c = c_1 \cdot c_2$.

Finally, based on the hyperbolic tangent function and the distance-reward ratio μ, the acceptance probability model is created by Eq. (13.7)

$$\theta = f(w, d) = \begin{cases} \text{th}(c \cdot \frac{w}{d}), & \text{if } w \geq w_0 \text{ and } d \leq MTD; \\ 0, & \text{otherwise.} \end{cases} \tag{13.7}$$

The acceptance probability, θ, belongs to $[0, 1]$ for all w and d. Moreover, $\theta = 0$ if either w or d does not meet its threshold. Moreover, the partial derivatives satisfy Eq. (13.8)

$$\frac{\partial f}{\partial w} > 0, \quad \frac{\partial f}{\partial d} > 0, \tag{13.8}$$

when $w \geq w_0$ and $d \leq MTD$. The proposed function f satisfies all the requirements put forward above, and therefore is suitable for simulating the relationships between reward, w, distance, d, and acceptance probability, θ.

With the definition of acceptance probability in Eq. (13.7), θ can be calculated given w and d, and the ASP can be evaluated according to θ.

The following assumptions are made: (a) in a circle region $\odot_r^t$ with radius, r, and the center, t, there are n workers; and (b) all the tasks and all the workers are independent from each other. Therefore, whether a task is accepted by a worker follows binomial distribution. Inspired by To et al. [216], given a task t in $\odot_r^t$, the proposed ASP probability model is defined by Eq. (13.9)

$$ASP = 1 - (1 - \theta)^n. \tag{13.9}$$

In Eq. (13.9), $1 - \theta$ represents the probability that a worker refuses t, $(1 - \theta)^n$ shows the probability that n workers in the circle refuse the task. Therefore, $1 - (1 - \theta)^n$ indicates the probability that at least one worker will accept the task, namely, the ASP of t.

13.3.3.2 Optimized Strategy and Radius Estimation

To assign tasks efficiently, we must obtain the minimum radius, r, for task, t. It is obvious that when the *ASP* of each task is equal to a given threshold E_{ASP}, the minimum radius, r, for t can be solved.

We summarize our proposed strategy in the following steps, then provide the details of each step with theoretical analyses:

1. Input E_{ASP} and a contour plot, evaluate the worker density den_i for each task t_i according to the task location and contour lines in contour plot;
2. Set $w_i = W/|T|$, compute r_i with Eq. (13.16) for the given tasks;
3. Reset $w_i = w_0 + \frac{r_i}{\sum_{i=1}^{|T|} r_i}(W - w_0|T|)$, where w_0 is the lower bound of reward. Then return to step 2 to re-compute r_i with the tuned w_i.

Specifically, the worker density is evaluated for each task in Step 1: if the location of t_i is on a contour line, the density is the value of the contour; when t_i is between two contour lines, the density is approximated by the mean of those two contour values. Step 2 calculates r_i with the mean reward. Step 3 tunes w_i for each task in proportion to the pre-computed r_i, namely, increase the reward for the tasks with large r_i while decrease the reward for the ones with small r_i. Then r_i is calculated again. Finally, the workers who are located in area $\odot_{r_i}^{t_i}$ will be informed to perform task t_i with the reward w_i.

Before solving Eq. (13.9), the SC-server has to measure the distance, d, and the number of the workers in $\odot_r^t$. Without additional prior knowledge, it is natural to assume all workers are distributed uniformly in $\odot_r^t$. Suppose worker, p, is at the coordinates (x, y), the expectation of $\sqrt{(x^2 + y^2)}$ can be considered with respect to d, i.e,

$$E(r) = \iint_{\odot_r^t} \sqrt{(x^2 + y^2)} \cdot \frac{1}{\pi r^2} \, dx \, dy = \frac{2}{3} r. \tag{13.10}$$

The total number of workers in an area is calculated by the worker density, *den*, in $\odot_r^t$, which is shown in Eq. (13.11).

$$n = \pi r^2 \cdot den, \tag{13.11}$$

where πr^2 is the area of $\odot_r^t$, as shown in Fig. 13.6.

Four equations, Eqs. (13.7) and (13.9)–(13.11), contain the relationships with reward, distance and acceptance probability. As the objective is to determine the radius, r, within which workers should be informed of each task, these equations are used to generate r.

Combining Eqs. (13.9) and (13.11) and eliminating n, we have

$$1 - E_{ASP} = (1 - \theta)^{\pi r^2 \cdot den}. \tag{13.12}$$

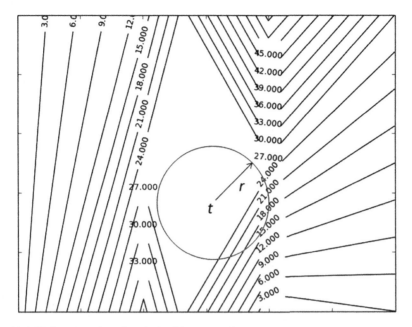

Fig. 13.6 Estimate number of workers with contour plot

Then, taking a logarithmic transformation on both side of Eq. (13.12), we have

$$\frac{\ln(1 - E_{ASP})}{\pi \cdot den} = r^2 \cdot \ln(1 - \theta). \tag{13.13}$$

Given we have modelled the acceptance probability, θ, as a function of reward, w, and distance, d, incorporating Eq. (13.7) into Eq. (13.13) and substituting d with Eq. (13.10), we have the following equation,

$$-\frac{\ln(1 - E_{ASP})}{\pi \cdot den} = \ln\left(\frac{e^{\frac{3cw}{r}} + 1}{2}\right). \tag{13.14}$$

Solving Eq. (13.14) gives radius, r. However, this equation is a transcendental equation, which means it cannot be solved through traditional mathematical deduction. Practically, there are two ways to solve transcendental equations. The first is to set an accuracy threshold and design a greedy algorithm to search a wide solution space. This type of method usually leads to expensive computational costs. Alternatively, Taylor's formula can approximate the right side of the equation denoted as $h(r)$, set $r_0 = 1$, and thus we have

$$h(r) \approx h(r_0) + h'(r_0)(r - r_0). \tag{13.15}$$

Therefore, after derivation, we have

$$-\frac{\ln(1 - E_{ASP})}{\pi \cdot den} \approx \left[2 \cdot \ln \frac{e^k + 1}{2} - k \cdot \frac{e^k}{e^k + 1} \right] \cdot r - \left[\ln \frac{e^k + 1}{2} - k \cdot \frac{e^k}{e^k + 1} \right],$$
(13.16)

where $k = 3cw$. Equation (13.16) is an affine function that can be solved.

In summary, the proposed method first builds a contour plot with full privacy budget, then models acceptance probability and ASP based on hyperbolic tangent function. Taylor's formula is used to solve the model's equations. Finally, the task-informing radius is calculated with an optimized-reward distribution, as opposed to the traditional uniform reward.

13.3.4 Experimental Evaluation

13.3.4.1 Performance on DE

The variation in the tendencies of DE for the Gowalla dataset, along with other parameters is shown in Fig. 13.7. Specifically, Fig. 13.7a shows that DE increases as the E_{ASP} increase, because achieving a larger E_{ASP} requires a larger number of workers participate in task assignment and thus leads to more distance error. In addition, the distance errors in the proposed optimized-reward method are generally less than those of the uniform-reward method, which indicates that the total distance error can be reduced by shrinking the radius distribution of the tasks in a smaller interval.

Figure 13.7b shows that the distance error decreases when the total reward is increased. Obviously, when a task's reward is much higher, the probability of a worker accepting it will be higher too. Thus, less workers are needed to achieve the E_{ASP}, and this leads to a small distance error. However, once the reward is increased to a certain level, the probability of a worker accepting the task becomes stable, leading to an approximately fixed number of workers who will accept the task. Thus the latter half of the curve flattens. Given difference between the amount of the

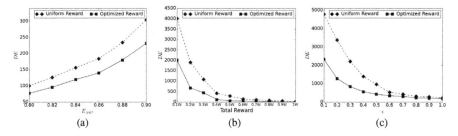

(a)   (b)   (c)

Fig. 13.7 DE estimation with Gowalla dataset. (**a**) DE VS E_{ASP}. (**b**) DE VS Reward. (**c**) DE VS ϵ

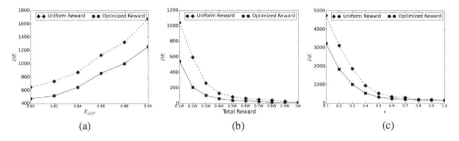

Fig. 13.8 *DE* estimation with T-driver dataset. (**a**) *DE* VS E_{ASP}. (**b**) *DE* VS Reward. (**c**) *DE* VS ϵ

total rewards, distance errors in the proposed optimized-reward method were always smaller than those of the uniform-reward methods.

Figure 13.7c illustrates the change in *DE* when the privacy budget varies. A smaller ϵ means more volume of noise is added to each cell, which leads to a larger distance error. Increasing the privacy budget, caused the volume of noise to decrease to a stable level, thus the distance error levels off at the end of the curve.

Experimental results conducted on the T-Driver dataset are shown in Fig. 13.8, where the total reward is $0.1W$ and $\epsilon = 0.3$ in Fig. 13.8a, $E_{ASP} = 0.88$ and $\epsilon = 0.4$ in Fig. 13.8b, $E_{ASP} = 0.9$, the total reward is $0.1W$ in Fig. 13.8c. The results show the same variation tendency of the *DE* as the other parameters, and demonstrates the reliability of the proposed method.

13.3.4.2 Performance on *RejR*

This experiment studies the *RejR* with different values for: E_{ASP}, the total reward and the privacy budget in optimized-reward and uniform-reward strategies, respectively. Results show that the optimized-reward method significantly outperformed the uniform-reward method against the metric, *RejR*.

Figure 13.9b shows results for the Gowalla dataset, where the total reward is $0.1W$ and privacy budget is 0.3. As shown in Fig. 13.9a, the average travel distance of each task, namely $2r/3$, was distributed in a wide interval when we applied the uniform-reward method, while the distribution was suppressed into a quite narrow range with the optimized-reward method. Therefore, when *MTD* is defined as the distance that 90% of workers are willing to travel, represented by the horizontal line in Fig. 13.9a. The *RejR* of the uniform-reward method was approximately 0.1, while it was 0 with the optimized-reward method. We achieved the same results when the value of E_{ASP} was varied from 0.80 to 0.90, as shown in Fig. 13.9b.

We also conducted the experiments with a different total reward and privacy budget, the results were almost the same as Fig. 13.9b. This demonstrates that the proposed optimized-reward method can efficiently decrease the average distance of a task by increasing its reward, thus ensuring that tasks with sparse workers can be accepted with a high probability.

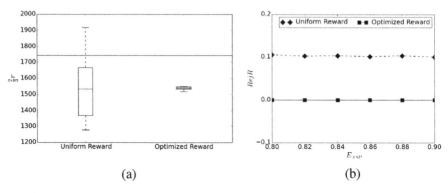

Fig. 13.9 $RejR$ with E_{ASP}. (**a**) The Boxplot of $\frac{2}{3}r$. (**b**) $RejR$ VS E_{ASP}

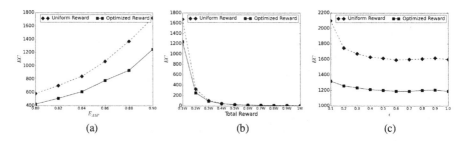

Fig. 13.10 EC estimation with Gowalla dataset. (**a**) EC VS E_{ASP}. (**b**) EC VS Reward. (**c**) EC VS ϵ

13.3.4.3   Performance on EC

Experimental results on the Gowalla dataset show that EC increases when the E_{ASP} is increased, and decreases when the total reward and the privacy budget is increased.

Figure 13.10a illustrates the relationship between EC and E_{ASP}, with a total reward of $0.1W$ and $\epsilon = 0.2$. The reason that EC increases with an increasing E_{ASP} is that a higher E_{ASP} is achieved by assigning the task to more workers which leads to a larger assignment radius. Given a fixed E_{ASP}, the radius generated with the proposed optimized-reward method varies within a small range, resulting in a smaller EC, compared to that of the uniform-reward method.

Figure 13.10b shows the value of EC with the total reward varying from $0.1W$ to W. If the total reward is doubled or tripled, from the beginning, the EC decreases significantly, because a higher reward, logically, increases the probability that a worker will accept the task. When the probability reaches a high enough level to become stable, the stimulating effect of reward gradually fades and the EC curve flattens.

We also investigated the behavior of EC with different values for ϵ, a fixed reward and a fixed E_{ASP}. As shown in Fig. 13.10c, when the E_{ASP} was set to 0.9, the total

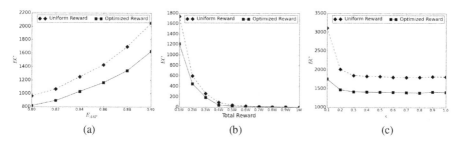

Fig. 13.11 EC estimation with T-driver dataset. (**a**) EC vs E_{ASP}. (**b**) EC vs Reward. (**c**) EC vs ϵ

reward was $0.1W$, and ϵ was increased from the beginning, the EC decreased but tended to become stable quickly. This implies that the total execution cost, as an aggregate metric, is insensitive to noise. The result also shows that the total EC caused by the proposed optimized-reward method is always smaller than that of the uniform-reward method, and that the EC can be diminished by tuning the reward allocation properly.

Experiments on the T-driver dataset are given in Fig. 13.11. They show that EC has similar variation tendency when the parameters are changed, which demonstrates the efficiency of the proposed method.

In the above discussions, we present and analyse the variation tendency with various E_{ASP}, W and ϵ. We conclude that optimized reward outperforms uniform reward in the execution cost.

13.4 Summary

Privacy issues are becoming increasingly concerning with the popularity of spatial crowdsourcing. The main challenge in applying differential privacy to spatial crowdsourcing is to achieve an optimal trade-off between privacy and effectiveness. This chapter present an existing method, trusted third party, to preserving privacy. Based on that, a reward-based SC is proposed to address this challenge. This method first constructs a contour plot with a differential privacy guarantee to minimize the magnitude of by fully using the given privacy budget so as to achieve accurate task assignments. Then two models to calculate the assignment success probability and the acceptance probability, respectively, are constructed to ensure efficient task arraignment. The method can dramatically enhance the task acceptance ratio through adjusting each task's reward. Future work will extend the proposed method to scenarios with redundant tasks assignments, and frameworks for SC without a trusted third party will be explored.

Chapter 14
Correlated Differential Privacy for Non-IID Datasets

14.1 Introduction

Although differential privacy has been widely accepted, previous work has mainly focused on independent datasets which assumes all records were sampled from a universe independently. Despite this, a real-world dataset often exhibits strong coupling relations: some records are often correlated with each other, and this may disclose more information than expected. For example, differential privacy ensures that deleting a user will not affect the aggregated result. However, in a social network dataset, users are always interacting with other users, and this kind of relationship may provide helpful information for identifying those deleted users. Another example assumes members in the same family may have a high probability of catching the same infectious disease. If an adversary knows one person gets the flu, he has a high probability of inferring the health of this person's family. We refer to this relationship as *correlated information*, and the involved records related to each other are *correlated records*. An adversary with knowledge on the correlated information will have a higher chance of obtaining private information [126], and violating the definition of differential privacy. Hence, how to preserve rigorous differential privacy in a correlated dataset is an emerging issue that needs to be addressed.

Over the last decade, limited research has been concerned with correlated differential privacy. A pioneer study by Kifer et al. [126], confirmed that if correlated records are ignored, the released data will have a lower than expected privacy guarantee. Their successive paper proposed a new privacy definition named *Pufferfish* [127], which takes the correlated records into consideration, but it does not meet the requirement of differential privacy. Chen et al. [41] dealt with the correlated problem in social networks by multiplying the original sensitivity with the number of correlated records. This straightforward method was not optimal because it introduced a large amount of noise into the output, that overwhelmed the true answer and demolished the utility of the dataset. Hence, a major research

© Springer International Publishing AG 2017
T. Zhu et al., *Differential Privacy and Applications*,
Advances in Information Security 69, DOI 10.1007/978-3-319-62004-6_14

barrier in correlated differential privacy is that the correlated dataset can provide extra information to the adversary, which can not be modeled by the traditional mechanism. In such a situation, satisfying the definition of differential privacy is a more complicated task.

As advances in correlated data analysis are made, especially with recent developments in the research of *non-iid* data [32], it is now possible to overcome the research barrier mentioned above. The correlated information can be modeled by functions or parameters that can be further defined as background information in the differential privacy mechanism. For example, Cao et al. [32] utilized the time interval and correlation analysis to identify correlated records and model correlated information by inter-behavior functions. This solution can help tackle the research barrier by incorporating the modeled information to the differential privacy mechanism.

However, there are still three main challenges with this approach:

- The first challenge is how to identify and represent correlated records. Records are often correlated in terms of certain relationships that are not obvious. A deep exploration of the relationship is necessary to understand which records are correlated and how they interact with others.
- The second challenge lies in the fact that if an algorithm just increases the sensitivity by multiplying it with the number of correlated records, the new sensitivity will be large, especially when lots of records couple with each other. To achieve a rigorous privacy guarantee, a large magnitude of noise has to be added, and this will significantly decrease the utility of the correlated dataset.
- The third challenge occurs when answering a large number of queries. When the number of queries is large, the privacy budget has to be divided into many small parts, which increases the noise for each query. This problem is more serious in a correlated dataset because the more queries that need to be answered, the more correlated records that will be involved, and the larger the amount of noise that will be introduced.

All these challenges imply correlated information should not be incorporated into differential privacy in a straight forward manner, and a novel mechanism is in high demand. This chapter first presents two basic definition on the correlated differential privacy: Pufferfish [127] and Blowfish [98], and then proposes a comprehensive correlated differential privacy solution, including sensitivity measurement and mechanism design. Table 14.1 shows the application setting for correlated differential privacy.

14.2 An Example: Correlated Records in a Dataset

Most existing differential privacy works assume the dataset consists of independent records. However, in real world applications, records are often correlated with each other. Kifer et al. [126] pointed out that differential privacy without considering

Table 14.1 Application settings

Application	Non-IID dataset publishing
Input data	Non-IID dataset
Output data	Count query answer
Publishing setting	Interactive
Challenges	Correlated information; large set of queries
Solutions	Adjust sensitivity measurement; iteration mechanism
Selected mechanism	Iteration
Utility measurement	Error measurement
Utility analysis	Laplace property
Privacy analysis	Sequential composition
Experimental evaluation	Performance measured by error distance

Table 14.2 Frequency dataset

Attribute	Count
A	2
B	100
C	200

correlation between records will decrease the privacy guarantee on the correlated dataset. For example, suppose a record r has influence on a set of other records, and this set of records will provide evidence on r even though record r is deleted from the dataset. In this scenario, traditional differential privacy fails to provide sufficient privacy as it claims.

Suppose we want to publish a dataset D with n records. To simplify the example, we assume there is only one attribute in D and its values are A, B and C. Dataset D can then be easily transferred to frequency dataset x in Table 14.2, where the *Attribute* column stores the attribute values and the *Count* column represents the number of records with each value. The target of privacy preserving is to hide the true count in x.

To preserve ϵ-differential privacy, the randomization mechanism M will add independent noise to the count. Since deleting a record will impact the count number at most by 1, the sensitivity of the count query is 1, and the independent noise will be sampled from the Laplace distribution $Laplace(\frac{1}{\epsilon})$.

This process works well when records are sampled independently from domain $\mathscr{X}$. However, if some records are correlated with each other, traditional differential privacy may under estimate the privacy risk [126]. For example, let the frequency dataset x in Table 14.2 represent a medical report in which *Attribute* represents the address and *Count* denotes the number of patients who have the Flu. Suppose a patient named *Alice* and her nine immediate family members are living at the same address B. When *Alice* contracts the Flu, the entire family will also be infected. In this case, deleting the record of *Alice* in address B will impact nine other records, and the count of address B will change to 90 (*Alice* got the Flu) or remain 100 (*Alice* is healthy). Suppose the noisy count returns 99 and the noise is sampled from

$Laplace(\frac{1}{\epsilon})$. This means there is high probability $Alice$ is healthy because the query answer is close to 100. Specifically, the answer 99 is $e^{10\epsilon}$ times more likely than the probability of $Alice$ to get the Flu. Compared to the independent records with privacy bounded in e^{ϵ}, correlated records have a probability of ten times more likely to be disclosed. In this instance, traditional differential privacy seriously mismatches the reality for correlated records.

We define the problem as a correlated differential privacy problem. To deal with this, one possible solution is to design a new sensitivity measurement based on the relationship between records. A naive way to measure the sensitivity is to multiply global sensitivity with the number of correlated records. In the above mentioned example, while deleting $Alice$ will impact at most ten records, the sensitivity is re-measured as 1×10, and the noise will be sampled from $Laplace(\frac{10}{\epsilon})$.

This naive sensitivity measurement can be extended to differential scenarios. For instance, if A, B, C in Table 14.2 are linear dependent, that is $A + B = C$, deleting a record in A will eliminate 1 count in A and 1 count in C at the most, and sensitivity will be measured to 2. If we have $A * B = C$, deleting a record in A will at most change the count of 100 in C, so the sensitivity is measured as $max(count(A), count(B))$. It is obvious that in some cases, sensitivities will be very high, leading to considerable redundance noise. This naive solution is not optimal in correlated differential privacy. How to define new sensitivity in a proper way is a problem of critical importance.

In summary, a major disadvantage of traditional differential privacy is overlooking the relationship between records, which means the query result leaks more information than is allowed. If we deal with the problem by simply multiplying the number of correlated records to the sensitivity, the query result will contain redundant noise and damages the utility of the dataset. Consequently, a sophisticated solution to the correlated differential privacy problem is urgently needed.

To deal with this problem, Kifer et al. successive paper proposed a new privacy framework, $Pufferfish$ [127], which allows application domain experts to add extra data relationships to develop a customized privacy definition.

14.3 Basic Methods

14.3.1 Pufferfish

A Pufferfish privacy framework has three components: set S of secrets, a set $Q \in S \times S$ of secret pairs, and a class of data distributions Θ, which controls the amount of allowable correlation in the data.

Secret S is a set of potential sensitive statements, for example, Alice is in the dataset D, and she got Flu. Discriminative pair Q are mutually exclusive pair of secrets such as Alice got Flu, Alice got diabetes. Each $\theta \in \Theta$ represents a belief that an adversary hold about the data. The selection of Θ is tricky. If it is too restrictive, the privacy may be not guaranteed sufficiently. If Θ is too broad, the privacy mechanisms will lead to little utility.

Definition 14.1 (Pufferfish Privacy) A privacy mechanism M is said to be ϵ-Pufferfish private with parameters S, Q and Θ if for all datasets D with distribution $\theta \in \Theta$, and for all secrete pairs $(s_i, s_j) \in Q$, and for all possible output ϕ, we have

$$e^{-\epsilon} \leq \frac{P(M(D) = \phi|s_i, \theta)}{P(M(D) = \phi|s_j, \theta)} \leq e^{\epsilon}, \tag{14.1}$$

where $P(s_i|\theta) \neq 0$, and $P(s_j|\theta) \neq 0$.

As Pufferfish privacy definition proposes Θ that can capture the correlation within dataset, the framework is used widely in the correlated differential privacy analysis. Differential privacy is a special case of Pufferfish, when every property about an individual's record in the data is kept secret, and Θ is a set of all distributions where each individuals private value is distributed independently (no correlation) [226].

14.3.2 Blowfish

Based on Pufferfish, He et al. [98] developed Blowfish framework to provide more parameters to curators, who can extend differential privacy using a *policy*. Policy specifies which information must be kept secret about individuals, and what constraints may be known publicly about the data.

The secret is defined by a discriminative secret graph $G(V, E)$, where V is the set of all values that an individual's value record can take, and E denotes the set of discriminative pairs.

Definition 14.2 (Policy) A policy is a triple $P(Q, G, I_C)$, where $G = (V, E)$ is a discriminative secret graph with $V \subseteq Q$. S is the set of secrets and the set of discriminative pairs is defined as $Q \in S \times S$. I_C denotes the set of databases that are possible under the constraints C that are known about the dataset.

Definition 14.3 (Blowfish Privacy) A privacy mechanism M is said to be ϵ, P-Blowfish private with $P(Q, G, I_C)$ if neighboring datasets (suppose x in D and y in D') in the constraint of P, and for all possible output ϕ, we have

$$e^{-\epsilon} \leq \frac{P(M(D) = \phi)}{P(M(D') = \phi)} \leq e^{\epsilon}, \tag{14.2}$$

where $P(s_i|\theta) \neq 0$, and $P(s_j|\theta) \neq 0$.

For any x and y in the domain, Eq. (14.3) can be interpreted as Eq. (14.3.2) as follows:

$$e^{-\epsilon \cdot d(x,y)} \le \frac{P(M(D) = \phi)}{P(M(D') = \phi)} \le e^{\epsilon \cdot d(x,y)}, \tag{14.3}$$

where $d(x, y)$ is the shortest distance between x and y in graph G. Adversary is allowed to distinguished between x and y that appear in different disconnected component in G.

When addressing the privacy issue in a correlated dataset, the essential problem is how to model the extra background information introduced by correlated records. This problems can be addressed with the development of correlated data analysis. Unlike pufferfish, Zhu et al. [259] proposed a practical way to achieve differential privacy with less noise. The solution will be investigated in the following sections. More specifically, the authors attempt to address the following research issues:

- How to identify correlated records in a dataset?
- How to calibrate the sensitivity for correlated records?
- How to design a correlated data publishing mechanism?

14.4   Correlated Differential Privacy

14.4.1   Correlated Differential Privacy Problem

For simplicity, *correlated*, *relationship* and *correlation* are interchangeable in this chapter. If a record $r_i \in D$ is correlated to other $k - 1$ records, this group of k ($k \le |D|$) records are called as *correlated records*, which is denoted by a set $\mathbf{r_i} = \{r_j \in D | \text{all } r_j \text{ are correlated to } r_i\}$. Dataset D is then referred to as a *correlated dataset*. The i.i.d dataset is a special case of a correlated dataset, in which $k = 1$. k varies from different datasets and is independent to queries. Query is denote by f and a group of queries is denoted as F. The set of records q that are related to a query f is referred to as the query's responding records.

If a correlated dataset D contains d attributes, it is more convenient to map D to a histogram x over domain $\mathcal{X}$. Each bin b represents the combination of attributes, and the number of bins is denoted by N. The frequencies in a histogram is the fraction of the count of bins, which are denoted by $x(b_i), (i \in N)$. For example, Table 14.2 is actually a histogram with bins A, B and C, whose frequency $x(A) = 0.0066$, $x(B) = 0.3311$ and $x(C) = 0.6623$, respectively. Formally, the histogram representation can be defined as follows:

Definition 14.4 (Histogram Representation)   A dataset D can be represented by a histogram x in a domain $\mathcal{X}: x \in N^{|\mathcal{X}|}$, Two datasets D and D' are defined as *neighboring datasets* if and only if their corresponding histograms x and x' satisfy $||x - x'||_1 \le 1$.

Another important notion is *correlated degree*. The naive multiple method assumes deleting a record will definitely change other records in a correlated dataset.

However, most records are only partially correlated, and deleting a record may have a different impacts on other records. These impacts is defined as the *correlated degree* of records.

Definition 14.5 (Correlated Degree)   Suppose two records r_i and r_j are correlated to each other. This means the relationship between them is represented by the *correlated degree* $\delta_{ij} \in [-1, 1]$ and $|\delta_{ij}| \geq \delta_0$, where δ_0 is the threshold of the correlated degree.

Corollary 14.1   *If $\delta_{ij} < 0$, r_i and r_j have a negative coupling; if $\delta_{ij} > 0$, they have a positive coupling; $\delta = 0$ indicates no relationship. If $|\delta_{ij}| = 1$, record r_i and r_j are fully correlated with each other.*

The correlated degree represents the impact of a record on another record. The smaller absolute value of δ_{ij} illustrates a *weak* coupling, and indicates that deleting r_i will have a low possibility of impacting r_j. When δ_{ij} is closed to 1 or -1, the coupling is strong, and deleting r_i will greatly impact r_j. However, in real world applications, few records are fully correlated, and this observation can be useful in our proposed method.

From the perspective correlated data analysis, it is possible to list all relationships between records and maintain a *correlated degree matrix* Δ, in which $\delta \in \Delta$.

$$\Delta = \begin{pmatrix} \delta_{11} & \delta_{12} & \dots & \delta_{1n} \\ \delta_{21} & \delta_{22} & \dots & \delta_{2n} \\ \dots & \dots & \dots & \dots \\ \delta_{n1} & \delta_{n2} & \dots & \delta_{nn} \end{pmatrix}. \tag{14.4}$$

Here are four properties of Δ: (1) It is *symmetrical* with $\delta_{ij} = \delta_{ji}$, which indicates the relationship between two records is irrelevant to their sequence; (2) Elements on the diagonal are equal to 1, which implies every record is fully correlated with itself; (3) A threshold δ_0 is defined to filter the weak correlated degree. In Δ, $|\delta_{ij}| \geq \delta_0$. If $|\delta_{ij}| < \delta_0$, δ_{ij} is set to zero; (4) It is *sparse*. Only parts of records are correlated with each other.

The above terms and correlated degree matrix will help solve the correlated differential privacy problem.

14.4.2   Research Issues and Challenges

Privacy preserving on a correlated dataset is challenging because of its special dataset structure and corresponding privacy requirement. Introducing differential privacy to a correlated dataset, brings three major challenges.

- *How to identify correlated records in a dateset?*
  It is often hard to identify correlated records and correlated degree δ. Different types of datasets may have various ways to couple with their records. Moreover,

several records may mix together and have exponential possible relationships, thus making correlated analysis very complex.

- *How to calibrate sensitivity for correlated records?*
  Traditional *global sensitivity* may not be suitable for correlated datasets due to large noise. In our previous Flu example, *global sensitivity* introduces ten times larger noise to the count output in a correlated dataset. In addition, local sensitivity can not be used because it still only relates to an individual record without considering coupling information.
- *How to re-design the differential privacy mechanism?*
  Even *correlated sensitivity* can significantly decrease noise compared to large noise when answering a large set of queries for *global sensitivity*. When dealing with the correlated dataset, the traditional mechanism may not be suitable for a correlated dataset. A new mechanism is expected to satisfy differential privacy, as well as retain sufficient utility for future applications.

14.4.3   Correlated Dataset Analysis

Several studies on correlated data analysis have attempted to identify and model correlated information. Correlated information can be identified by correlated analysis including *time interval analysis*, *attribute analysis*, or *similarity analysis*. Cao et al. [33] presented a correlated *Hidden Markov* detection model to detect abnormal group-based trading behaviors. They defined a time interval and assumed behaviors falling into the same interval as correlated behaviors. Song et al. [204] proposed a hybrid coupling framework, which applied some particular attributes to identify relationships among records. Zhang et al. [250] identified the network traffic correlated record using an IP address. They presented a correlated network traffic classification algorithm.

Correlated information can be modeled in varies ways. Cao et al. [33] modeled correlated information using the inter-couple and intra-couple behavior functions. These functions were adopted in the correlated framework to represent the correlated degree between behaviors. Zhou et al. [255] mapped the correlated records to an undirected graph and proposed the multi-instance learning algorithm.

These approaches help to model background information for differential privacy. An advanced differential privacy releasing mechanism will be proposed with the aim of guaranteeing a sufficient privacy level as well as decreasing extra noise.

Correlated analysis is carried out to generate the correlated degree matrix Δ for a correlated dataset. This can be done in various ways depending on the background knowledge of the curator or the characteristics of the dataset. Typical methods can be conceptualized into two categories.

The first type of correlated analysis assumes the curator or the attacker obtained the background knowledge in advance. The correlated degree matrix Δ is pre-defined as background knowledge. Taking Table 14.2 as an example. The curator or attacker discover there are full coupling relationships among *A*, *B* and *C*, e.g.

$A + B = C$ or $A * B = C$. Δ can then be created according to the background information. Identifying a full coupling relationship among records is relatively easy. But for some weak couplings, they needs further domain knowledge or determination by an expert.

Another type of correlated analysis can be carried out without any direct background knowledge. The correlated degree will be defined in various ways.

1. *Attribute analysis.* This utilizes certain particular attributes to discover the relationships among records. When the values of these attributes are the same or similar to each other, records with those values are considered as correlated records. For example, the `address` attribute can be used to determine family members in a survey dataset. In a network traffic dataset, the `IP-address` attribute can help identify traffic coming from the same host. Moreover, the similarity in attribute values can be adopted to measure the correlated degree; a high similarity implies a strong coupling. This method can identify correlated records effectively and accurately. However, it can hardly be implemented when no attribute is available to disclose the relationship.
2. *Time interval analysis.* This method pre-defines the size of a time interval to identify the correlation in the stream dataset. Records falling into the same interval are considered as correlated records. For instance, Cao et al. [33] aggregated the behaviours within time intervals and modeled the coupling between these activities. This method can figure out the multiple records mixed together but is only suitable for a time related dataset.
3. *Pearson Correlation analysis.* If the dataset contains no proper attribute or time information to identify the correlated information, the *Pearson Correlation Coefficient* is an efficient way to discover correlated records. It extracts all or parts of an attribute in a correlated dataset and calculates the *Pearson Correlation Coefficient* between records. By defining the threshold δ_0, the correlated degree matrix can be generated according to the correlation coefficient. Other correlation or distance measurements can also be applied. For example, Song et al. [204] applied KL divergence to measure the correlated degree between records. However, this type of method can only identify the linear correlation between records.

Other strategies also exist for correlated analysis. However, no matter what methods are applied, the target is to define the correlated degree matrix Δ, which plays an essential role in correlated differential privacy.

14.4.4   Correlated Sensitivity

Traditional *global sensitivity* will result in redundant noise derived from both records and queries. For the record, as analyzed earlier, the traditional method assumes records are fully correlated with each other, and therefore, it just multiplies the *global sensitivity* with the maximal number of correlated records leading to large noise. For a query, the traditional method uses a fixed *global sensitive* without considering the prosperity of different queries. In actual fact, only some of the responding records are correlated with others, and the curator only needs to consider the correlated information within these responding records. Hence, sensitivity should be adaptive for both the correlated record and the query.

Based on this observation, *correlated sensitivity* can defined, which takes both record and query into consideration. The notion of *record sensitivity* is relating to the correlated degree of each record. Based on this notion, the *correlated sensitivity* associated with the query is proposed.

Definition 14.6 (Record Sensitivity)   For a given Δ and a query f, the record sensitivity of r_i is

$$CS_i = \sum_{j=0}^{n} |\delta_{ij}|(\||f(D^j) - f(D^{-j})\||_1), \qquad (14.5)$$

where $\delta_{ij} \in \Delta$.

The *record sensitivity* measures the effect on all records in D when deleting a record r_i. $\delta_{ij} \in \Delta$ estimates the correlated degree between records r_i and $r_j \in D$. This notion combines the number of correlated records and the correlated degree together. If D is an independent dataset, CS_i is equal to the global sensitivity.

Definition 14.7 (Correlated Sensitivity)   For a query f, *correlated sensitivity* is determined by the maximal record sensitivity,

$$CS_q = \max_{i \in q}(CS_i), \qquad (14.6)$$

where q is a record set of all records responding to a query f.

Correlated sensitivity is related to a query f. It lists all the records q responding to f and selects the maximal *record sensitivity* as the *correlated sensitivity*. The advantage of the measurement is that when a query only covers the independent or weak correlated record, correlated sensitivity will not bring extra noise.

After defining *correlated sensitivity* for each query f, the noisy answer will eventually be calibrated by the following equation:

$$\widehat{f}(D) = f(D) + Laplace\left(\frac{CS_q}{\epsilon}\right). \qquad (14.7)$$

Correlated sensitivity CS_q will be smaller than *global sensitivity* GS and *local sensitivity* LS. Both assume each record is fully correlated with each other and the correlated degree is also ignored.

Lemma 14.1   *For a query f, correlated sensitivity is equal to or less than the* global sensitivity GS and the local sensitivity LS.

Proof Suppose there are at most k correlated records in a dataset D, then we have $GS = k \cdot \max_{D,D'}(||f(D) - f(D')||_1)$, and $CS_i = \sum_{j=1}^{n} \delta_{ij}(||f(D^j) - f(D^{-j})||_1)$. Because at most k records are correlated, we have $\sum_{j=1}^{n} \delta_{ij} = \sum_{j=1}^{k} \delta_{ij} \leq k$. As $||f(D^j) - f(D^{-j})||_1 \leq \max_{D,D'}(||f(D) - f(D')||_1)$, we have $CS_i \leq GS$. As any CS_i are less or equal to GS, for a query f, we have $CS_q \leq GS$.

For the *local sensitivity* $LS = k \cdot \max_{D'}(||f(D) - f(D')||_1)$, we also have $||f(D^j) - f(D^{-j})||_1 \leq \max_{D'}(||f(D) - f(D')||_1)$ and $\sum_{j=1}^{n} \delta_{ij} = \sum_{j=1}^{k} \delta_{ij} \leq k$, then we have $CS_q \leq LS$.

Correlated sensitivity CS can be used in various types of data releasing mechanisms. If records in the dataset are independent, the *CS* will be equal to the global sensitivity, while for the correlated dataset, the *CS* will introduce less noise than *GS* and *LS*.

14.4.5 Correlated Iteration Mechanism

Even though *correlated sensitivity* decreases the noise compared with *global sensitivity*, when dealing with a large number of queries, the answers still have high noise because the privacy budget has to be divided into several small parts. This is especially so when the records are strongly correlated with others and the noise is significantly higher than the independent dataset. To tackle the problem, an iterative-based mechanism will be adopted to limit the noise in the query answer.

The iterative-based mechanism was first proposed by Hardt et al. [92] who constructed a dataset sequence to answer all queries by iteratively updating the datasets. When a given query witnesses a significant difference between the current dataset and the true dataset, the mechanism updates the current dataset in the next iteration [92]. The main advantage of this mechanism is that it can save the privacy budget and decrease the noise when confronting lots of queries. Hence, it will be suitable for data releasing in the correlated dataset.

In this section, a **Coupled Iteration Mechanism** (*CIM*) is proposed to answer a set of queries on the correlated dataset.

14.4.5.1 Overview of Correlated Iteration Mechanism

The *CIM* aims to release the results of a set of queries by iteratively updating the dataset under the notion of differential privacy. In this procedure, a dataset is represented by a histogram x with length N. Let t be the round index, and the histogram be represented by x_t at the end of round t. The curator is given a query set F and select a f_t in each round t. Let a_t denotes the true answer and $\hat{a}_t$ denotes the noisy answer:

$$a_t = f_t(x), \tag{14.8}$$

$$\widehat{a}_t = f_t(x) + Laplace\left(\frac{CS_{q_t}}{\epsilon}\right). \tag{14.9}$$

The difference between the true answer given by x_{t-1} and the noisy answer from x_t is denoted by $\widehat{d}_t$:

$$\widehat{d}_t = f_t(x_{t-1}) - \widehat{a}_t. \tag{14.10}$$

This is utilized to control the update round in each iteration. At a high level, *CIM* maintains a sequence of histogram x_0, x_1,..., x_t, which gives increasing approximation to the original dataset x.

The mechanism is shown in Algorithm 1. Firstly, the privacy budget is divided into several parts and the histogram is initialized as the uniform distribution x_0. In each round t, curators select a query $f_t \in F$, using x_t to generate the answer $a_t = f_t(x_t)$ and the noise answer $\widehat{a}_t$. The distance $\widehat{d}_t$ between the query f_t on x_{t-1} and the noisy answer $\widehat{a}_t$ is computed. If $|\widehat{d}_t|$ is less than a threshold T, the x_{t-1} is considered to be a good approximation of x on query f_t. Curators will release the $f_t(x_{t-1})$ directly and put the x_{t-1} into the next iteration. If the distance is larger than the threshold, the histogram x_{t-1} will be improved in this round. Curators will release $\widehat{a}_t$ and use an correlated updating function U to generate the new histogram x_t.

The *CIM* aims to answer a large group of queries with limited privacy budgets on a correlated dataset. In summary, this mechanism has the following features:

- First, it takes the relationship between records into consideration. It applies not only *correlated sensitivity*, but more importantly, it develops a correlated update function to improve the histogram in each iteration.
- Second, it decreases the total amount of noise. The *CIM* maintains a histogram sequence to answer a set of queries F, rather than using a single histogram to answer all queries. One histogram in the sequence roughly corresponds to one query in F. This way, each histogram can approximate the close answer to the true answer.
- Finally, more queries can be answered than the traditional mechanism with the same privacy budget. Only the update steps will consume the privacy budget. Algorithm 1 indicates that even for a very large set of queries, the number of update rounds is still less than the total number of queries.

14.4.5.2 Correlated Update Function

This section defines a correlated update function U in the histogram context. For a histogram x_{t-1}, the function U firstly identifies all responded records $r \in q_t$. For each record in q_t, all correlated records are listed and denoted as superset $\mathbf{q_t}$. The update function U then identifies a set of bins $\mathbf{b}$ that contain $\mathbf{q_t}$ and re-arranges the

---

Algorithm 1 Correlated Iteration Mechanism

Require: $x, \epsilon, F = f_1, ..., f_L, L, \Delta, T$.
Ensure: $F(x)$
   1. $\epsilon_0 = \frac{\epsilon \eta^2 \delta_0^2}{\log N}$;
  for $i = 1, ..., N$ **do**
    2. $x_0(b_i) = 1/N$;
  end for
  for each round $t \leftarrow 1...L$ **do**
    3. select a query f_t;
    4. sample λ_t from $Laplace(CS_{q_t}/\epsilon_0)$;
    5. compute the noise answer $\widehat{a}_t = f_t(x) + \lambda_t$;
    6. compute $\widehat{d}_t = f_t(x_{t-1}) - \widehat{a}_t$;
   if $|\widehat{d}_t| \leq T$ **then**
     7. $x_t = x_{t-1}$, output $f_t(x_{t-1})$;
     8. continue;
   else
     9. $x_t = U(x_{t-1})$, output $\widehat{a}_t$;
   end if
  end for

Algorithm 2 Correlated Update Function

Require: $x_{t-1}, \widehat{d}, f_t, \Delta, \eta$.
Ensure: x_t.
   1. Identifying q_t;
   2. Identifying the correlated record set $\mathbf{q_t}$;
   3. Identifying the bin set $\mathbf{b}$ contains $\mathbf{q_t}$;
  for For each bin $b_i \in \mathbf{b}$ **do**
    4. Update the frequency of $x(b_i)$;
  end for
   5. Normalization of x_t

---

frequency of each bin in $\mathbf{b}$ according to Eq. (14.11). The final frequency of the x_t will be normalized so they sum to 1.

Definition 14.8 (Correlated Update Function) Let $x_0, x_2, ..., x_t$ be a histogram sequence, and function U is defined as a correlated update function if it satisfies $x_t = U(x_{t-1})$. The U is defined as:

$$x_t(b_i) = x_{t-1}(b_i) \cdot \exp(-\eta \cdot \delta_{q_t} \cdot y_t(x_{t-1})), \tag{14.11}$$

where $y_t(x_{t-1}) = f_t(x_{t-1})$ if $\widehat{d} > 0$ and otherwise, $y_t(x_{t-1}) = 1 - f_t(x_{t-1})$. η is an update parameter associated with the number of maximal update rounds.
Algorithm 2 shows the detailed procedure.

    The correlated update function is based on the intuition that if the answer derived from x_{t-1} is too small compared with the true answer, the frequency of the relative bins will be enhanced. Otherwise, curators will decrease the frequency if the answer is too large.

14.4.5.3 Parameters Discussion

This section discuss the estimation of parameters in *CIM*. As mentioned earlier, only the update round consumes the privacy budgets. To measure the parameters T and η, curators need to estimate the maximal number of update rounds u_{max} and the possible number of update rounds u_F. The u_{max} helps determine the privacy budgets in each iteration. In addition, the u_F for F is related to the accuracy.

First, curators measure the maximal number of update rounds u_{max}. Given a dataset x, the u_{max} can be measured based on the following lemma.

Lemma 14.2 *Given a histogram x with length N, the u_{max} for correlated update function U, defined by Eq.(14.12) is*

$$u_{max} = \frac{\log N}{\eta^2 \cdot \delta_0^2}. \tag{14.12}$$

Proof Give the original histogram x and the initial update histogram x_0, the *CIM* will update the x_0 in each round t until $x_t = x$. The u_{max} depends on how many steps that x_0 can be transferred to x. The method follows the update strategy of Hardt et al. [92], who define the distance between x_0 and x in terms of *relative potential*:

$$\phi_t = RE(x||x_t) = \sum x(b) \log \left(\frac{x(b)}{x_t(b)} \right). \tag{14.13}$$

Based on Eq.(14.13), $\phi_0 \leq \log N$. When ϕ_t drops to zero, the update will be terminated and $x_t = x$. According to Hardt et al. [92], the potential drops in each round is at least $\eta^2 \delta_0^2$, therefore there are at most $\frac{\log N}{\eta^2 \cdot \delta_0^2}$ rounds in the *CIM*.

The u_{max} is utilized to determine the privacy budget ϵ_0 in each round. Equation (14.14) shows the calculation of ϵ_0:

$$\epsilon_0 = \frac{\eta^2 \delta^2 \epsilon}{\log N}. \tag{14.14}$$

Compared to the traditional data releasing mechanism, which divides the privacy budget ϵ according to the number of queries, the algorithm can easily demonstrate that $\epsilon_0 \geq \epsilon/|F|$.

Lemma 14.2 also indicates u_{max} is associated with parameter η and the couple parameter threshold δ_0. If the curator wants to successfully answer more queries, he/she can choose a smaller η to allow more rounds. However, this will lead to larger noise in each query answer because the privacy budget ϵ_0 in each round will also be diminished.

Second, to estimate the possible number of update rounds u_F for a query set F, let the probability of updating be ρ_1 and the probability of non-updating be ρ_2, the algorithm has the follow Lemma:

Lemma 14.3 *When both the privacy budget ϵ_0 in each round and the parameter T are fixed, the probability of the update will be*

$$\rho_1 = \exp\left(\frac{-\epsilon_0|T - \alpha|}{CS_q}\right), \tag{14.15}$$

and the probability of the non-update will be

$$\rho_2 = 1 - \exp\left(\frac{-\epsilon_0|T - \alpha|}{CS_q}\right), \tag{14.16}$$

where α bounds the accuracy of the CIM.

Corollary 14.2 *Given a query set F, the u_F will satisfy Eq. (14.17)*

$$u_F = |f|\exp\left(\frac{-\epsilon_0|T - \alpha|}{CS_q}\right). \tag{14.17}$$

Proof Suppose we have $|F|$ queries, and altogether $|F|$ rounds. The probability of the update is $Pr(|\widehat{d_t}| > T)$. We have

$$Pr(|\widehat{d_t}| > T) = Pr(|f_t(x) + \lambda_t - f_t(x_{t-1})| > T).$$

Let $|f_t(x) - f_t(x_{t-1})| \leq \alpha$, and λ_t be sampled from *Laplace*(ϵ_0/CS_q) according to the property of Laplace distribution:

$$Pr(|\gamma| > t) = Pr(\gamma > t) + Pr(\gamma < t) = 2\int_t^\infty x\exp\left(-\frac{x}{\sigma}\right)dx = \quad (14.18)$$

$$= \exp\left(-\frac{t}{\sigma}\right) \quad (14.19)$$

Because $\sigma = \frac{CS_q}{\epsilon_0}$, we have

$$Pr(|\widehat{d_t}| > T) \leq \exp\left(\frac{\epsilon_0(\alpha - T)}{CS_q}\right).$$

If there are $|F|$ queries, the algorithm will update at most $|F|\exp\left(\frac{\epsilon_0(\alpha-T)}{CS_q}CS_q\right)$ rounds.

Lemma 14.3 shows the probability of the update is related to T and α. If parameter T is much smaller than α, the update probability will be very high and the noise will increase simultaneously which will affect the accuracy of the answer. However, if T is very large, even though we decrease the number of update rounds, but the output answer is always far away from the true answer, which also decreases the accuracy, we can conclude the accuracy of *CIM* is related to T. Section 14.4.7 uses the experiment to demonstrate the trade-off between T and the accuracy of *CIM*.

14.4.6 Mechanism Analysis

The proposed *CIM* aims to obtain an acceptable utility with a fixed ϵ differentially privacy budget. In this section, we will first prove the algorithm is satisfied with ϵ-differential privacy, and then analyze the utility cost.

14.4.6.1 Privacy Analysis

To prove *CIM* is satisfied with differential privacy, one needs to analyze which steps in *CIM* will consume the privacy budget. Algorithm 1, accesses the histogram and generate a noisy answer in each round. However, the noisy answer is only used to check whether the current histogram is accurate enough to answer the query. In most rounds, the algorithm does not release the noisy answer, and therefore the algorithm consumes no privacy budget. The noisy answer is only released in the update round when the current histogram is not accurate enough. Consequently, the privacy budget is only consumed in the update step and the privacy analysis can be easily limited in the correlated update functions.

The sequential composition accumulates privacy budget ϵ for each step when a series of private analyses is performed sequentially on the dataset. As mentioned earlier, given a x, we have $u_{max} = \eta^{-2}\delta^{-2}\log N$. The privacy budget ϵ_0 allocated to each round is $\epsilon_0 = \frac{\epsilon\eta^2\delta_0^2}{\log N}$. According to the sequential composition, the released answers for the query set F will consume the $\epsilon_0 * u_F$ privacy budget. Because $u_F \leq u_{max}$, we have $\epsilon_0 * u_F \leq \epsilon$. Consequently, the proposed *CIM* algorithm preserves ϵ-differential privacy.

14.4.6.2 Utility Analysis

For the utility analysis, error measurement is used. In *CIM*, the utility is measured by a set of released query answers. Accordingly, the error is measured by the maximal distance between the original and the noisy answer.

Definition 14.9 ((α,β)-Accuracy for *CIM*) The *CIM* is (α,β)-accuracy for a set of query F, if: with probability $1 - \beta$, for every query $f \in F$ and every x, we have

$$\max_{f \in F, t \in L} |CIM_t(x_t) - f_t(x)| \leq \alpha, \qquad (14.20)$$

where $CIM_t(x_t)$ is the output of *CIM* in round t.

Based on the definition, we will demonstrate *CIM* mechanism is bounded by a certain value α with a high probability.

Theorem 14.1 *For any query $f \in F$, for all $\beta > 0$, with probability at least $1 - \beta$, the error of* CIM *output is less than α. When*

$$\alpha \geq \frac{CS_q}{2\epsilon_0} \left(\log \frac{\rho_1 \rho_2 L}{\beta} \right) + \frac{T}{2},$$

the CIM is satisfied with (α, β)-accuracy.

Proof The value of $CIM_t(x_t)$ is determined by the $\widehat{d}$, which results in the non-update round or the update round. Both scenarios will be considered in the utility measurement. Let $error_{non\text{-}update}$ represents the error introduced by non-update rounds and $error_{update}$ denotes the error introduced by update rounds. According to the union bound, we have

$$Pr(\max_{f \in F, t \in L} |CIM_t(x_t) - f_t(x)| > \alpha) \leq \rho_1 * Pr(error_{non\text{-}update} > \alpha)$$

$$+ \rho_2 * Pr(error_{update} > \alpha).$$

If $\widehat{d} \leq T$, it will be a non-update round and *CIM* will output $f_t(x_{t-1})$.

$$error_{non\text{-}update} = |CIM(x_t) - f_t(x)| = |f_t(x_{t-1}) - f_t(x)|.$$

Because

$$|\widehat{d}_t| = |f_t(x_{t-1}) - \widehat{a}_t| \leq T,$$

we have

$$|f_t(x_{t-1}) - f_t(x)| \leq T + \lambda_t,$$

where $\lambda_t \sim Laplace\left(\frac{CS_{q_t}}{\epsilon'}\right)$. According to the property of Laplace distribution $Laplace(b)$, we have

$$Pr(error_{non\text{-}update} > \alpha) = Pr(\max_{t \in L} |T + \lambda| > \alpha) \leq L * Pr(|T + \lambda| > \alpha),$$

and

$$\leq L \exp\left(\frac{-|\alpha - T|\epsilon_0}{CS_q}\right),$$

where $CS_q = \max_{t \in 1, \dots, m} CS_{q_t}$.

If $\widehat{d} > T$, it will be an update round and *CIM* will output $\widehat{a}_t$. We have

$$error_{update} = |CIM(x_t) - f_t(x)| = |f_t(x) + \lambda_t - f_t(x)| = |\lambda_t|.$$

Then we have

$$Pr(error_{update} > \alpha) = Pr(\max_{t \in L} |\lambda| > \alpha) \leq L * Pr(|\lambda| > \alpha),$$

and

$$\leq L \exp\left(\frac{-\alpha\epsilon_0}{CS_q}\right).$$

Accordingly,

$$Pr(\max_{f\in F, t\in L} |CIM_t(x_t) - f_t(x)| > \alpha) \leq L\rho_1 \exp\left(\frac{-|\alpha - T|\epsilon_0}{CS_q}\right) + L\rho_2 \exp\left(\frac{-\alpha\epsilon_0}{CS_q}\right).$$

Let

$$L\rho_1 \exp\left(\frac{-|\alpha - T|\epsilon_0}{CS_q}\right) + L\rho_2 \exp\left(\frac{-\alpha\epsilon_0}{CS_q}\right) \leq \beta,$$

we have

$$\rho_1 \exp\left(\frac{-|\alpha - T|\epsilon_0}{CS_q}\right) + \rho_2 \exp\left(\frac{-\alpha\epsilon_0}{CS_q}\right) \leq \frac{\beta}{L}$$

$$\Rightarrow \log \rho_1\rho_2 + \frac{(T - 2\alpha)\epsilon_0}{CS_q} \leq \log\frac{\beta}{L}$$

$$\Rightarrow \alpha \geq \frac{CS_q}{2\epsilon_0}\left(\log\frac{\rho_1\rho_2 L}{\beta}\right) + \frac{T}{2}.$$

14.4.7 Experiment and Analysis

14.4.7.1 Datasets and Configuration

The experiments involve six datasets:

Adult   The Adult dataset from the *UCI Machine Learning repository*[1] originally had 48,842 records and 14 attributes. After deleting the missing records
and filtering the attributes, the experiment eventually had 30,162 records with 15
dimensions.

IDS   This dataset was collected by *The Third International Knowledge Discovery
and Data Mining Tools Competition* that aimed to build an *Intrusion Detection
System* (IDS). The experiment sampled a subset with 40,124 records and 65
dimensions.

---

[1] http://archive.ics.uci.edu/ml/.

NLTCS   The *National Long-Term Case Study* (NLTCS) dataset was derived from *StatLib*.[2] It contained 21,574 records with 16 binary attributes, which corresponded to 16 functional disability measures. These are six activities of daily living and ten instrumental activities of daily living.

Three other datasets were derived from Hay's work [96], which have been widely used in the differentially private histogram release test.

Search Logs   This synthetic data set was generated by interpolating *Google Trends* data and *America Online* search logs. It contained 32,768 records, each of which stored the frequency of searches with the keyword *Obama* within a 90 min interval between *January 1, 2004* and *August 9, 2009*.

NetTrace   This contained the IP-level network trace at a border gateway of a university. Each record reported the number of external hosts connected to an internal host. There were 65,536 records with connection numbers ranging from 1 to 1423.

Social Network   This records the friendship relations among 11,000 students from the same institution, sampled from an online social network web site. Each record contains the number of friends of certain students. There were 32,768 students, each of who had at most 1678 friends.

All datasets contained no pre-defined correlated information. The *Pearson Correlation Coefficient* is used to generate the correlated degree matrix Δ with the threshold $\delta_0 = 0.6$. Approximately all datasets had a quarter of their records correlated with some other records, and the maximal size of the correlated group was around 10. For each dataset, a query set F was generated with 10,000 random linear queries and each answer fell into $[0, 1]$. The accuracy of results was measured by *Mean Square Error* (MSE).

$$MSE = \frac{1}{|F|} \sum_{f_i \in F}^{F} (\widehat{f_i}(x) - f_i(x))^2. \tag{14.21}$$

A lower MSE implies a better utility for the corresponding mechanism.

14.4.7.2   The Performance of *CIM*

The performance of the *CIM* mechanism was examined in this section through comparison with the state-of-the-art naive *Laplace Mechanism* (LM) [62]. To show its effectiveness, *correlated sensitivity* (CS) in both *CIM* and *LM* was tested, and are denoted as CIM&CS and LM&CS, respectively. The experiments were conducted on all datasets and the privacy budgets varied from 0.1 to 1. Two parameters, T and η, were set to 0.3000 and 7.0000, respectively.

---

[2]http://lib.stat.cmu.edu/.

As shown in Fig. 14.1, *CIM* has lower MSE than *LM* on all datasets. Specifically, for the Adult dataset in Fig. 14.1a, when $\epsilon = 0.4$, the *CIM* achieves a MSE of 0.3593 while *LM* achieves 0.5171. Thus, *CIM* outperfroms *LM* by 43.84%. When $\epsilon = 1$, *CIM* achieves a MSE of 0.0491 which outperformed *LM* by 73.12%. These results illustrate that in correlated datasets, *CIM* outperforms *LM* when answering a large set of queries. The improvement by *CIM* can also be observed in Fig. 14.1b–f. The proposed *CIM* has better performance because it only consumes the privacy budget in the update rounds, which is less than the total number of queries $|F|$. While the traditional *LM* mechanism consumes the privacy budget when answering every query, this actually leads to inaccurate answers. The experimental results show the effectiveness of *CIM* when answering a large set of queries.

In the context of differential privacy, the privacy budget ϵ serves as a key parameter to determine privacy. Figure 14.1 shows the impact of ϵ on the performance of *CIM*. According to Dwork [62], $\epsilon = 1$ or less would be suitable for privacy preserving purposes. For a comprehensive investigation, the *CIM*'s performance is evaluated under various privacy preserving levels by varying the privacy budget ϵ from 0.1 to 1 with a 0.1 step on six datasets. It was observed that as ϵ increases, the MSE evaluation becomes better, which means the lower the privacy preserving level, the larger the utility. In Fig. 14.1a, the MSE of *CIM* is 5.5350 when $\epsilon = 0.1$. Even though it preserves a strict privacy guarantee, the query answer is quite inaccurate. When $\epsilon = 0.7$, the MSE drops to 0.1025, retaining an acceptable utility of the result. The same trend can be observed on other datasets. For example, when $\epsilon = 0.7$, the MSE is 0.1894 in Fig. 14.1b, and is 0.1733 in Fig. 14.1c. These results confirm the utility will be enhanced as the privacy budget increases.

Moreover, it is observed that the MSE decreased much faster when ϵ ascends from 0.1 to 0.4 compared to when ϵ ascends from 0.4 to 1. This indicates a larger utility cost is needed to achieve a higher privacy level ($\epsilon = 0.1$). It is observed that the performance for both the *CIM* and *LM* mechanism was stable when $\epsilon \geq 0.7$. This indicates the *CIM* was capable of retaining the utility for data releasing while satisfying a suitable privacy preserving requirement.

In addition, the MSE of *CIM* with the non-privacy release is compared. If the curator answers all queries without any privacy guarantee, the MSE is 0. Figure 14.1 shows the MSE of *CIM* was very close to 0 when $\epsilon \geq 0.7$. This was because *CIM* applied iterative steps and *correlated sensitivity* to reduce the magnitude of introduced noise. This result confirms correlated differential privacy can ensure rigorous privacy with an acceptable utility loss.

The evaluation shows the effectiveness of *CIM* on several aspects.

1. The proposed *CIM* can retain a higher utility of released data compared with the *LM*.
2. As the privacy budget increased, the performance of *CIM* was significantly enhanced. A suitable privacy guarantee can be selected to achieve a better tradeoff.
3. When we have a sufficient privacy budget, the utility loss of released data is small.

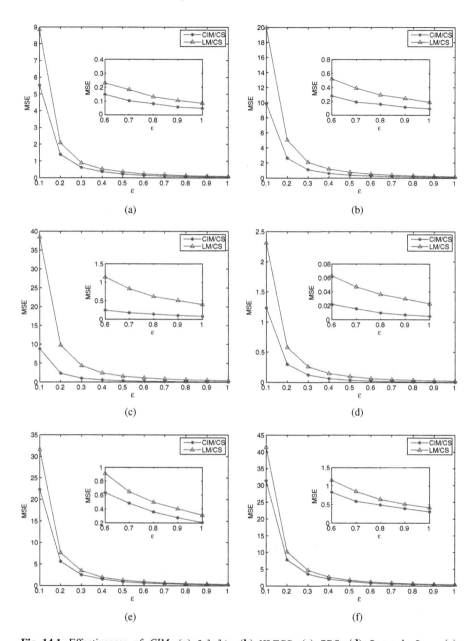

Fig. 14.1 Effectiveness of *CIM*. (**a**) Adult. (**b**) NLTCS. (**c**) IDS. (**d**) Search Log. (**e**) NetTrace. (**f**) Social Network

Table 14.3 The maximal
size of correlated groups

Datasets	k
Adult	15
NLTCS	10
IDS	10
Search Log	10
Net Trace	10
Social Network	20

14.4.7.3 Correlated Sensitivity vs. Global Sensitivity

In this subsection, the performance of the *correlated sensitivity* is measured. In order to show the effectiveness of *correlated sensitivity*, the experiment selected the *LM* to answer the query set F and compared its performance of *correlated sensitivity* with *global sensitivity* (GS). To deal with the correlated dataset, *global sensitivity* was multiplied by k, the maximal number of records in a correlated group. Table 14.3 lists the size k of different datasets.

Figure 14.2 shows the results in which the sensitivities are termed CS and GS, with the privacy budget varying from 0.1 to 1.0. It can be observed that all MSE measures of *correlated sensitivity* on all six datasets were less than MSE of *global sensitivity* with different privacy budgets. These results imply *correlated sensitivity* leads to less error than *global sensitivity* in the context of correlated datesets. Specifically, as shown in Fig. 14.2a, when $\epsilon = 1$, the *LM* with CS achieves an MSE at 0.0850. This outperforms the *LM* result with a GS of 0.2293. The performance of improvement is more significant as the privacy budget decreases. When $\epsilon = 0.3$, the MSE of CS is 0.8906, which is much lower than MSE of the *LM* with a GS of 2.4785.

Moreover, it is clear MSE of *correlated sensitivity* is close to 0, which indicates *correlated sensitivity* can achieve the privacy preserving purpose while retaining a high utility of query answers.

Similar trends can be observed on other datasets as shown in Fig. 14.2b–f. For example, in Fig. 14.2b, the *LM* with CS leads to an MSE of 0.5222, which is much smaller than 3.1534 from the *LM* with GS. These results illustrate CS is independent to the characteristics of datasets.

The experimental results illustrate that even with the naive mechanism, *correlated sensitivity* can lead to a better performance than *global sensitivity*. It confirms the conclusion in Lemma 14.1, which proves the magnitude of *correlated sensitivity* is equal to or less than the *global sensitivity*.

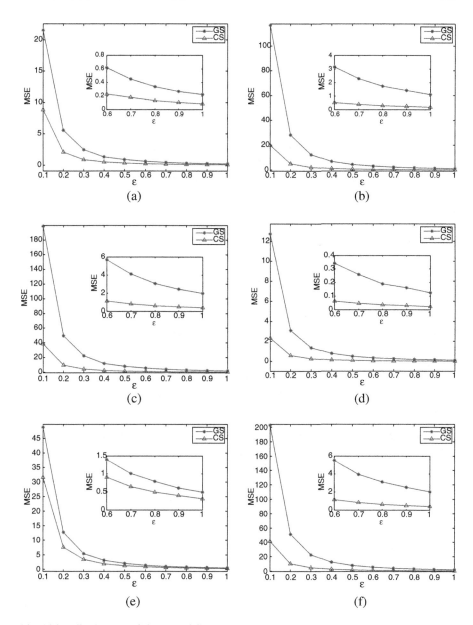

Fig. 14.2 Effectiveness of *Correlated Sensitivity*. (**a**) Adult. (**b**) NLTCS. (**c**) IDS. (**d**) Search Log. (**e**) NetTrace. (**f**) Social Network

14.5 Summary

Traditional differential privacy mainly focuses on independent datasets, failing to offer sufficient privacy guarantees for correlated datasets. Solving the correlated differential privacy problem is challenging as we need to find out the correlated records, measure the correlated sensitivity, and re-design the differential privacy release mechanism to answer a large number of queries. The correlated differential privacy solution presented in this chapter addressed these challenges effectively. It first categorizes possible correlated record analysis methods for different types of datasets, then the definition and analysis of correlated sensitivity are presented to measure the sensitivity of the query on the correlated records. Finally an effective correlated data releasing mechanism is described to enhance performance when answering a large group of queries. This chapter proposes two popular definitions, pufferfish and blowfish, on correlated dataset, and proposes a solution to correlated differential privacy with the following contributions:

- The problem of correlated differential privacy is identified. The correlated problem can be extended to more complex applications, such as a correlated private recommender system or correlated location privacy.
- Novel correlated sensitivity is proposed to deal with correlated records. Compared to global sensitivity, it guarantees more rigorous privacy while retaining an acceptable utility.
- A novel data releasing mechanism is proposed to enhance performance when answering a large group of queries. This mechanism has proven to ensure better query results than Laplace mechanism.

These provide a practical way to apply a correlated privacy notion to differential privacy with less utility loss. The experimental results also show the robustness and effectiveness of the proposed solution.

Chapter 15
Future Directions and Conclusion

While previous chapters provide a thorough description on differential privacy and presents several applications in reality, many interesting and promising issues remain unexplored. The development of social networks provides great opportunities for research on privacy-preserving but also presents a challenge in effective utilization of the large volume of data. There are still other topics that need to be considered in differential privacy, and we consider a few directions that are worthy of future attention.

15.1 Adaptive Data Analysis: Generalization in Machine Learning

Adaptive data analysis creates a connection between differential privacy and machine learning. The machine learning theory has been well developed based on the assumption that a learning algorithm operates on a freshly sampled dataset. However, data samples are often reused and the practice of learning is naturally adaptive. This adaptivity breaks the standard generalization guarantees in machine learning and in the theory of hypothesis testing. It is easy to overfit the data when the adaptive data analysis is done using direct access to the dataset.

The property of differential privacy can partly solve this adaptive problem. Dwork et al. [65, 66] proposed the technique for performing arbitrary adaptive data analyses together with rigorous generalization guarantees. The authors claimed that differential privacy is an algorithmic stability guarantee, and algorithmic stability is known to prevent over-fitting. They provided an error bound quantitatively in terms of generality.

© Springer International Publishing AG 2017
T. Zhu et al., *Differential Privacy and Applications*,
Advances in Information Security 69, DOI 10.1007/978-3-319-62004-6_15

This line of research proves that the differential privacy has a generalization property that can be applied in statistical analysis and machine learning [16, 174] though at this early stage, many new possibilities on generalization need to be explored.

15.2   Personalized Privacy

It is common that data owners have quite different expectations regarding the acceptable level of privacy for their data. Consequently, differential privacy may lead to insufficient privacy protection for some users, while over-protecting others. If we can relax the level of privacy for some data owners, a higher level of utility can often be achieved. To this extend, a personalized differential privacy is required, in which users can specify a personal privacy for their data [111].

Ebadi [76] proposed a *Personalized Differential Privacy* (PDP) framework that arranges various privacy budgets to each record. When a query is performed on the dataset, the privacy level will be determined by the summary of privacy budgets of all responding records. Alaggan et al. [10] considered the privacy expectation not only for the individual user, but also for the individual item. They introduced a concept of heterogeneous differential privacy as opposed to previous models that implicitly assume uniform privacy requirements. Koufogiannis [129] introduced a situation of releasing sensitivity when the privacy level is subject to change over time. Its intuition can guide the privacy level allocation on personalized privacy problem.

These are tentative frameworks on personalized privacy, but how to determine the privacy budget for each record or individual still need to be tackled.

15.3   Secure Multiparty Computations with Differential Privacy

Most existing work is concerned with the centralized model of differential privacy, in which a trusted curator holds the entire private dataset, and computes it in a differentially private way. If a dataset is divided among multiple curators who are mutually untrusting each other, however, how can they compute differentially private messages for communication between themselves? Mironov et al. [158] explored a two-party scenario by showing a lower bound for the problem of computing the hamming distance between two datasets. Mohammed et al. [159] presented a two-party protocol based on the exponential mechanism. Both solutions are unlikely to be valid when the number of curator's lies is more than two.

How to preserve distributed differential privacy within multiple parties is a future topic for open-ended theoretical exploration. The solution will involve with inter-discipline techniques, including privacy preserving, security protocol designing and cryptography.

15.4 Differential Privacy and Mechanism Design

Mechanism design is the field of algorithm design where the inputs to the mechanism are controlled by strategic agents who may manipulate their inputs [15]. In this setting, the design of the mechanism must convince agents to provide their correct inputs to the mechanism.

A typical application of mechanism design is the auction design. Considering a scenario in which a data analyst wishes to buy information from a population to estimate some statistical information, while the owners of the private data experience some cost for their loss of privacy, agents between the sellers and the buyers wish to maximize their profit, so the goal is to design a truthful auction mechanism while preserving the privacy of the dataset.

Differential privacy limits any individual's influence on the result. If the mechanism satisfies differential privacy, agents will have little incentive to deviate from truthful behavior since they can only change the selected equilibria to a small degree.

McSherry [157] first proposed designing auction mechanisms using differentially private mechanisms as the building blocks. This private mechanism is only approximately truthful. Nissim et al. [173] showed how to convert differentially private mechanisms into exactly truthful mechanisms. Cummings et al. [51] studied the multi-dimensional aggregative games and solved the equilibrium selection problem. Barthe et al. [15] introduced a relational refinement type system for verifying mechanism design and differential privacy.

These series of research work focus on the mechanism design based on the differential privacy, which take advantage of the property of differential privacy. The direction lies at the intersection of differential privacy, mechanism design and probabilistic programming languages.

15.5 Differential Privacy in Genetic Data

As the advances of Genome sequencing technology, highly detailed genetic data is being generated inexpensively at exponential rates. The collection and analysis of such data has the potential to accelerate biomedical discoveries, and to support various applications, including personalized medical services. Despite all the benefits, the broad dissemination of genetic data has major implications on personal privacy. According to Erlich and Narayanan [77], the privacy of sensitive information was ranked as one of their top concerns and a major determinant of participation in a study. The privacy issues associated with genetic data are complex because of its wide uses as well as information on more than just the individual from which the data was derived. Erlich and Narayanan [77] analysed the privacy breaching techniques that involve data mining and combining distinct resources to gain private information that is relevant to DNA data, and further categorized them into identity tracing attack, attribute disclosure attacks using DNA (ADAD) and completion attack. Naveed et al. [168] reviewed the mitigating strategies for such attacks, as well as contextualizing these attacks from the perspective of medicine and public policy.

In the context of *Genome-wide association studies* (GWAS), several initial studies have explored the capability of differential privacy methods in the release of statistics for GWAS data, or shifting the original locations of variants. Johnson and Shmatikov [110] developed a framework for ab initio exploration of case-control GWAS that provide privacy-preserving answers to key GWAS queries. They designed the operators to differentially private output the number of SNPs associated with the disease, the location of the most significant SNPs, and the *p*-values for any statistical test between a given SNP and the disease, etc. By considering the protection against set membership disclosure, Tramer et al. proposed a relaxation of the adversarial model of differential privacy, and shown that this weaker setting achieves higher utility [220].

However, currently the differential private data release is still impractical, because it introduces a large amount of noise even for a few *singe-nucleotide polymorphism* locations (SNPs) in a given population. It is uncertain whether there is a calibrated noise adding mechanism for GWAS data, which satisfies the requirement of differential privacy [77].

15.6   Local Differential Privacy

As we mentioned in Chap. 1, the privacy model can be inserted between trusted curators and public users. This scenario is defined as centralized differential privacy, in which the differential privacy mechanism performs on the centralized data before sharing with public users. When the privacy model is inserted between data contributors and untrusted curators, the differential privacy mechanism perform on individual data before submitting to curators. This is because existing privacy-preserving strategies such as differential privacy and secure multiparty computation are less reliable in a distributed context for the following reason: In the centralized settings, the data curator is considered as a trusted agent who has the full access to all users' data. However in the distributed setting, data acquisition is more like conducting a questionnaire survey, and the data curator should be considered as an un-trusted investigator who can request responses from users but without free access to all data. Although centralized mechanism such as differential privacy can be adopted here to design secure and privacy-preserving data sharing framework such as *DistDiffGen* [160], *CELS* [101], or *DP-SUBN* [209], the extensive time and computation cost makes it impractical when encountering big data.

Recently, the *Local Differential Privacy* (LDP) model has been proposed to address the above issue in an effective and efficient way. As shown in Fig. 15.1, in the centralized settings, the trust data curator acquires and aggregate all the accurate user data, and then employs the DP mechanisms to sanitize the data for public sharing. In the local setting, by contrast, the un-trusted data curator acquires all the

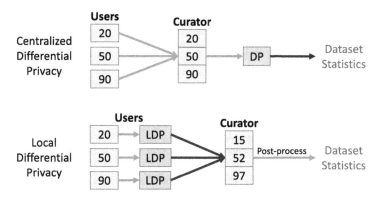

Fig. 15.1 Comparing LDP with DP models

noisy user data, then conducts post-process on them to obtain acceptable statistics which can be further shared with the public. Herein the statistics types which are available for the data curator to obtain would be limited to the population statistics and depend on both the design of local perturbation mechanism and post-process mechanism. That is to say, with the support of LDP model, the un-trusted data curator can estimate the approximate information of all users such as the prevalence rate in a population without inferring that which user has suffered from this sensitive disease.

Local privacy model was first formalized in [115], and then a theoretic upper bound under the LDP model was given by Duchi et al. [60]. Typically two main research questions are investigated:

1. How to design acceptable LDP mechanisms for different original data types generated by distributed users? For example, the ranges of data types can be from the single attribute data to multiple attribute data and even the set-valued data.
2. How to design acceptable LDP mechanisms that can achieve different analysis targets such as value estimation, distribution estimation and more complex machine learning tasks?

In the local differential privacy setting, the local perturbation mechanism conducted by each user should be a type of non-trivial differentially private noisy adding mechanism, such that it provides the data curator the ability to estimate certain statistics of the user population while satisfying differential privacy. In existing related research works, the non-trivial differentially private noisy adding mechanisms referred above are all based on *Randomized Response* technique. In the past several years, related LDP work have been on all three aspects of data life cycle: the data types existed in user node (*Original Data Type*), the data types submitted to data curator by users (*Uploaded Data Type*) and the targets that we want to achieve via analysing the uploaded data (*Post-process Target*). Table 15.1 gives an overview of above three aspects.

Table 15.1 Three aspects of data life cycle in local differential privacy

Original data type	Uploaded data type	Post-process target
Single numeric or categorical attribute	Noisy attribute value	Value estimation [230]; distribution estimation [78]
Multiple numeric or categorical attribute	Noisy attributes values	Value estimation [230]; mean and frequency estimation, complex machine learning tasks [171]; multi-dimensional joint distribution estimation [192]
Set-valued data	A random bit of noisy set-valued data [187]; noisy set-valued data [223]	Frequent item set estimation [187]; discrete distribution estimation [223]
Encoding data	A random bit of the encoding of location	Count estimation [42]

15.7 Learning Model Publishing

The difficulty in data publishing lies on the high correlation when meeting with large set of queries [102]. High correlation between queries leads to large volume of noise. According to the definition of sensitivity, correlations between m queries lead to higher sensitivity (normally m multiplied by the original sensitivity) than independent queries. The noise calibrated by this higher sensitivity will be added to each query answer. The accuracy of the results will be dramatically decreased compared to independent queries [140].

Current solutions aim to break the correlation by using transformation mechanism or publishing a synthetic dataset, both of which have been discussed in Chap. 5: Non-interactive setting. However, both solutions can only partly solve the problem, and at the same time, a new challenge is arising with the non-interactive setting: how to deal with unknown fresh queries. As the curator cannot know what users will ask after the data has been published, he/she has to consider all possible queries and adds pre-defined noise. When the dimension of the dataset is high, it is impossible to list all queries. Even if the curator is able to list all queries, this pre-defined noise will dramatically decrease the utility of the publishing results.

Zhu et al. [257] observed that these two challenges can be overcome by transferring the data publishing problem to a machine learning problem. They treated the queries as training samples which are used to generate a prediction model, rather than publishing a set of queries or a synthetic dataset. For correlations between queries, they apply limited queries to train the model. These limited queries have lower correlation than in the original query set. The model is also used to predict the remaining queries, including those fresh queries. Consequently, the model publishing method uses a limited number of queries to train an approximate accurate model and answers other fresh queries.

There are several advantages to publish a machine learning model:

- Many machine learning models can be applied to the data publishing problem. For example, we can use *linear regression, SVM for regression* and *neural network* to learn a suitable model M according to the training set. As prediction is a mature area that has been investigated in machine learning for several decades, we can choose sophisticated technologies and adjust parameters to obtain a better performance. In addition, lots of machine learning theories can be applied in this process.
- Some existing methods can be considered as an extension of model publishing. For example, the Private Multiplicative Weights (PMW) mechanism [92] is one of the most prevalent publishing methods in differential privacy. To some extent, it can be considered as an instance of the model publishing method. In the PMW, the histogram is a selected model and frequencies in this histogram constitute parameters of the model. The model (histogram) is trained by the input queries until it converges or meets the halting criteria. Compared to model publishing, however, PMW can only answer queries in the training set.
- The noisy model naturally has the property of generalization. Generalization is an essential problem in machine learning, but the differential privacy mechanism has proven that it can avoid over-fitting in the learning process [66].

The model publishing method is highly related to machine learning algorithms, but is different to private learning that mentioned in Chap. 6. Private learning introduces noise into the original learning algorithms, so that the privacy of the training dataset can be preserved in the learning process [115]. First, the purpose of publishing models is different. Model publishing method aims to publish a model for fresh query prediction, whereas private learning is only used for traditional machine learning tasks and will not preserve the privacy of fresh samples.

Second, the model publishing method considers pre-defined queries as a training set while private learning considers records in the original dataset as training samples. The target of differential privacy is to hide the true value of query answers, not the records, so the model publishing method considers the query as the training sample and the model is used to predict query answers rather than the values of records. In this respect, the model publishing method is totally different from private learning algorithms. Even though Kasiviswanathan [115] proved that Kearn's statistical query (SQ) [120] model can be implemented in a differentially private manner, the training set still comprises records in the dataset. Consequently, the *SQ* model is similar to private learning, not the model publishing method.

Finally, as public users normally use count, average or sum query, the model publishing method normally applies regression algorithms for true value prediction, while the private learning algorithm is usually specific to classification with labels of 0 or 1. Table 15.2 summarizes the major differences between the model publishing method and private learning.

With the large set of queries requirement of various applications, the model publishing method might be a promising method to integrate differential privacy with diverse applications.

Table 15.2 The difference between the model publishing method and private learning

	Model publishing method	Private learning
Model	The model is used to predict fresh query answers for public users	The model is used for traditional machine learning
Training set	Queries on the dataset	Records in the dataset
Protect target	Preserve the privacy of all queries	Do not protect future samples
Learning algorithms	Prediction	Classification

15.8   Conclusion

This book attempted a multi-disciplinary theories as well as applications of work on differential privacy. We provided an overview of the huge literature on two major directions: data publishing and analysis. For data publishing we have identified different publishing mechanisms and compared them in various natures of data. For data analysis, we discussed about two basic frameworks and illustrated their different analysis scenarios. The basic technique in differential privacy looks simple and intuitively appealing. When combining with specific problems, it is powerful and has been shown to be useful for diverse applications.

In the application part, we first present several major challenges, such as high sensitivity, sparsity of datasets, correlated data, etc., and then provide several solutions on how to solve those problems. Each application includes its background, and differentially private basic methods for this applications, and finally, we present a full solution in details.

Differential privacy still has lots of unknown potentials, and literature summarized in this book can be a starting point for exploring new challenges in the future. Our goal is to give an overview of existing works on differential privacy to show their usefulness to the newcomers as well as practitioners in various fields. We also hope the overview can help avoiding some redundant, ad hoc effort, both from researchers and from industries.

References

1. Differential privacy for everyone. download.microsoft.com/download/D/1/F/D1F0DFF5-8BA9-4BDF-8924-7816932F6825/Differential_Privacy_for_Everyone.pdf, 2012.
2. Bose sued over alleged privacy breach. https://www.itnews.com.au, 2017.
3. Telstra breaches privacy of thousands of customers. http://www.smh.com.au/it-pro/security-it, 2017.
4. Yahoo says 500 million accounts stolen. http://money.cnn.com/2016/09/22/technology/yahoo-data-breach/, 2017.
5. M. Abadi, A. Chu, I. Goodfellow, H. B. McMahan, I. Mironov, K. Talwar, and L. Zhang. Deep learning with differential privacy. In *CCS*, pages 308–318, 2016.
6. M. Abadi, A. Chu, I. J. Goodfellow, H. B. McMahan, I. Mironov, K. Talwar, and L. Zhang. Deep learning with differential privacy. In *Proceedings of the 2016 ACM SIGSAC Conference on Computer and Communications Security, Vienna, Austria, October 24–28, 2016*, pages 308–318, 2016.
7. G. Adomavicius and A. Tuzhilin. Toward the next generation of recommender systems: a survey of the state-of-the-art and possible extensions. *Knowledge and Data Engineering, IEEE Transactions on*, 17(6):734–749, June 2005.
8. C. C. Aggarwal and P. S. Yu, editors. *Privacy-Preserving Data Mining - Models and Algorithms*, volume 34 of *Advances in Database Systems*. Springer, 2008.
9. B. Agir, T. G. Papaioannou, R. Narendula, K. Aberer, and J.-P. Hubaux. User-side adaptive protection of location privacy in participatory sensing. *Geoinformatica*, 18(1):165–191, 2014.
10. M. Alaggan, S. Gambs, and A. Kermarrec. Heterogeneous differential privacy. *CoRR*, abs/1504.06998, 2015.
11. J. Anderson. *Hyperbolic Geometry*. Springer, second edition, 2005.
12. M. E. Andrés, N. E. Bordenabe, K. Chatzikokolakis, and C. Palamidessi. Geo-indistinguishability: Differential privacy for location-based systems. In *Proceedings of the 2013 ACM SIGSAC Conference on Computer; Communications Security*, CCS '13, pages 901–914, New York, NY, USA, 2013. ACM.
13. L. Backstrom, C. Dwork, and J. Kleinberg. Wherefore art thou r3579x?: Anonymized social networks, hidden patterns, and structural steganography. In *WWW'07*, pages 181–190, 2007.
14. B. Barak, K. Chaudhuri, C. Dwork, S. Kale, F. McSherry, and K. Talwar. Privacy, accuracy, and consistency too: a holistic solution to contingency table release. In *PODS '07*, pages 273–282, 2007.
15. G. Barthe, M. Gaboardi, E. J. Gallego Arias, J. Hsu, A. Roth, and P.-Y. Strub. Higher-order approximate relational refinement types for mechanism design and differential privacy. In *POPL*, pages 55–68, 2015.

© Springer International Publishing AG 2017   223
T. Zhu et al., *Differential Privacy and Applications*,
Advances in Information Security 69, DOI 10.1007/978-3-319-62004-6

16. R. Bassily, A. Smith, T. Steinke, and J. Ullman. More general queries and less generalization error in adaptive data analysis. *CoRR*, abs/1503.04843, 2015.

17. R. Bassily, A. D. Smith, and A. Thakurta. Private empirical risk minimization: Efficient algorithms and tight error bounds. In *FOCS*, pages 464–473, 2014.

18. A. Beimel, S. P. Kasiviswanathan, and K. Nissim. Bounds on the sample complexity for private learning and private data release. In *TCC*, pages 437–454. 2010.

19. A. Beimel, K. Nissim, and U. Stemmer. Characterizing the sample complexity of private learners. In *ITCS*, pages 97–110, 2013.

20. A. Beimel, K. Nissim, and U. Stemmer. Private learning and sanitization: Pure vs. approximate differential privacy. *CoRR*, abs/1407.2674, 2014.

21. A. Beimel, K. Nissim, and U. Stemmer. Learning privately with labeled and unlabeled examples. In *SODA*, pages 461–477, 2015.

22. Y. Bengio. Learning deep architectures for AI. *Foundations and Trends in Machine Learning*, 2(1):1–127, 2009.

23. S. Berkovsky, Y. Eytani, T. Kuflik, and F. Ricci. Enhancing privacy and preserving accuracy of a distributed collaborative filtering. RecSys '07, pages 9–16, New York, NY, USA, 2007. ACM.

24. R. Bhaskar, S. Laxman, A. Smith, and A. Thakurta. Discovering frequent patterns in sensitive data. In *SIGKDD*, pages 503–512, 2010.

25. D. M. Blei. Probabilistic topic models. *Commun. ACM*, 55(4):77–84, Apr. 2012.

26. D. M. Blei, A. Y. Ng, and M. I. Jordan. Latent Dirichlet allocation. *The Journal of Machine Learning Research*, 3:993–1022, Mar. 2003.

27. J. Blocki, A. Blum, A. Datta, and O. Sheffet. Differentially private data analysis of social networks via restricted sensitivity. In *ITCS*, pages 87–96, 2013.

28. A. Blum, C. Dwork, F. McSherry, and K. Nissim. Practical privacy: the SuLQ framework. In *PODS*, pages 128–138, 2005.

29. A. Blum, K. Ligett, and A. Roth. A learning theory approach to non-interactive database privacy. In *STOC*, pages 609–618, 2008.

30. J. A. Calandrino, A. Kilzer, A. Narayanan, E. W. Felten, and V. Shmatikov. "you might also like: " privacy risks of collaborative filtering. In *SP'11*, pages 231–246, 2011.

31. J. Canny. Collaborative filtering with privacy via factor analysis. SIGIR '02, pages 238–245, New York, NY, USA, 2002. ACM.

32. L. Cao. Non-iidness learning in behavioral and social data. *The Computer Journal*, 2013.

33. L. Cao, Y. Ou, and P. S. Yu. Coupled behavior analysis with applications. *IEEE Transactions on Knowledge and Data Engineering*, 24(8):1378–1392, 2012.

34. T.-H. H. Chan, E. Shi, and D. Song. Private and continual release of statistics. *ACM Trans. Inf. Syst. Secur.*, 14(3):26:1–26:24, 2011.

35. K. Chandrasekaran, J. Thaler, J. Ullman, and A. Wan. Faster private release of marginals on small databases. In *ITCS'14*, pages 387–402, 2014.

36. K. Chatzikokolakis, C. Palamidessi, and M. Stronati. Location privacy via geo-indistinguishability. *ACM SIGLOG News*, 2(3):46–69, 2015.

37. K. Chaudhuri and D. Hsu. Sample complexity bounds for differentially private learning. In *COLT*, pages 155–186, 2011.

38. K. Chaudhuri and C. Monteleoni. Privacy-preserving logistic regression. In *NIPS 2014*, pages 289–296, 2008.

39. K. Chaudhuri, C. Monteleoni, and A. D. Sarwate. Differentially private empirical risk minimization. *Journal of Machine Learning Research*, 12(2):1069–1109, 2011.

40. R. Chen, B. C. Fung, B. C. Desai, and N. M. Sossou. Differentially private transit data publication: a case study on the Montreal transportation system. In *SIGKDD*, pages 213–221, 2012.

41. R. Chen, B. C. Fung, S. Y. Philip, and B. C. Desai. Correlated network data publication via differential privacy. *The VLDB Journal*, 23(4):653–676, 2014.

42. R. Chen, H. Li, A. K. Qin, S. P. Kasiviswanathan, and H. Jin. Private spatial data aggregation in the local setting. In *ICDE*, pages 289–300, 2016.

43. R. Chen, N. Mohammed, B. C. M. Fung, B. C. Desai, and L. Xiong. Publishing set-valued data via differential privacy. *PVLDB*, 4(11):1087–1098, 2011.
44. R. Chen, Q. Xiao, Y. Zhang, and J. Xu. Differentially private high-dimensional data publication via sampling-based inference. In *SIGKDD*, pages 129–138, 2015.
45. S. Chen and S. Zhou. Recursive mechanism: Towards node differential privacy and unrestricted joins. In *SIGMOD*, pages 653–664, 2013.
46. C.-Y. Chow, M. F. Mokbel, and X. Liu. Spatial cloaking for anonymous location-based services in mobile peer-to-peer environments. *GeoInformatica*, 15(2):351–380, 2011.
47. CIFAR-10 and C.-. datasets. www.cs.toronto.edu/ kriz/cifar.html.
48. R. Collobert, K. Kavukcuoglu, and C. Farabet. Torch7: A matlab-like environment for machine learning. In *BigLearn, NIPS Workshop*, 2011.
49. G. Cormode, C. Procopiuc, D. Srivastava, E. Shen, and T. Yu. Differentially private spatial decompositions. In *ICDE*, pages 20–31, April 2012.
50. G. Cormode, D. Srivastava, N. Li, and T. Li. Minimizing minimality and maximizing utility: analyzing method-based attacks on anonymized data. *Proc. VLDB Endow.*, 3:1045–1056, September 2010.
51. R. Cummings, M. Kearns, A. Roth, and Z. S. Wu. Privacy and truthful equilibrium selection for aggregative games. *CoRR*, abs/1407.7740, 2014.
52. K.-H. Dang and K.-T. Cao. Towards reward-based spatial crowdsourcing. In *Control, Automation and Information Sciences (ICCAIS), 2013 International Conference on*, pages 363–368. IEEE, 2013.
53. W. Day, N. Li, and M. Lyu. Publishing graph degree distribution with node differential privacy. In *SIGMOD*, pages 123–138, 2016.
54. Y.-A. De Montjoye, C. A. Hidalgo, M. Verleysen, and V. D. Blondel. Unique in the crowd: The privacy bounds of human mobility. *Scientific reports*, 3:1376, 2013.
55. J. Dean, G. Corrado, R. Monga, K. Chen, M. Devin, Q. V. Le, M. Z. Mao, M. Ranzato, A. W. Senior, P. A. Tucker, K. Yang, and A. Y. Ng. Large scale distributed deep networks. In *NIPS*, pages 1232–1240, 2012.
56. R. Dewri. Local differential perturbations: Location privacy under approximate knowledge attackers. *IEEE Trans. Mob. Comput.*, 12(12):2360–2372, 2013.
57. R. Dewri. Local differential perturbations: Location privacy under approximate knowledge attackers. *IEEE Transactions on Mobile Computing*, 12(12):2360–2372, Dec. 2013.
58. B. Ding, M. Winslett, J. Han, and Z. Li. Differentially private data cubes: Optimizing noise sources and consistency. pages 217–228, 2011.
59. I. Dinur and K. Nissim. Revealing information while preserving privacy. In *PODS*, pages 202–210, 2003.
60. J. C. Duchi, M. I. Jordan, and M. J. Wainwright. Local privacy and statistical minimax rates. In *FOCS*, pages 429–438, 2013.
61. C. Dwork. Differential privacy. In *ICALP*, pages 1–12, 2006.
62. C. Dwork. Differential privacy: a survey of results. In *TAMC'08*, pages 1–19, 2008.
63. C. Dwork. Differential privacy in new settings. In *SODA '10*, pages 174–183, Philadelphia, PA, USA, 2010. Society for Industrial and Applied Mathematics.
64. C. Dwork. A firm foundation for private data analysis. *Commun. ACM*, 54(1):86–95, 2011.
65. C. Dwork, V. Feldman, M. Hardt, T. Pitassi, O. Reingold, and A. Roth. Generalization in adaptive data analysis and holdout reuse. In *NIPS*, pages 2350–2358, 2015.
66. C. Dwork, V. Feldman, M. Hardt, T. Pitassi, O. Reingold, and A. L. Roth. Preserving statistical validity in adaptive data analysis. In *STOC*, pages 117–126, 2015.
67. C. Dwork, K. Kenthapadi, F. McSherry, I. Mironov, and M. Naor. Our data, ourselves: Privacy via distributed noise generation. In *EUROCRYPT*, pages 486–503, 2006.
68. C. Dwork, F. McSherry, K. Nissim, and A. D. Smith. Calibrating noise to sensitivity in private data analysis. In *TCC*, pages 265–284, 2006.
69. C. Dwork, M. Naor, T. Pitassi, and G. N. Rothblum. Differential privacy under continual observation. In *STOC*, pages 715–724, 2010.

70. C. Dwork, M. Naor, T. Pitassi, G. N. Rothblum, and S. Yekhanin. Pan-private streaming algorithms. In *Innovations in Computer Science*, pages 66–80, 2010.
71. C. Dwork, M. Naor, O. Reingold, G. N. Rothblum, and S. Vadhan. On the complexity of differentially private data release: efficient algorithms and hardness results. In *STOC*, pages 381–390, 2009.
72. C. Dwork and A. Roth. The algorithmic foundations of differential privacy. *Found. Trends Theor. Comput. Sci.*, 9:211–407, Aug. 2014.
73. C. Dwork and A. Roth. The algorithmic foundations of differential privacy. *Found. Trends Theor. Comput. Sci.*, 9(3–4):211–407, Aug. 2014.
74. C. Dwork, G. N. Rothblum, and S. Vadhan. Boosting and differential privacy. In *2010 IEEE 51st Annual Symposium on Foundations of Computer Science*, pages 51–60, Oct 2010.
75. C. Dwork, A. Smith, T. Steinke, and J. Ullman. Exposed! a survey of attacks on private data. *Annual Review of Statistics and Its Application*, (0), 2017.
76. H. Ebadi, D. Sands, and G. Schneider. Differential privacy: Now it's getting personal. In *POPL*, pages 69–81, 2015.
77. Y. Erlich and A. Narayanan. Routes for breaching and protecting genetic privacy. *Nature Review Genetics*, 15:409–421, 2014.
78. G. C. Fanti, V. Pihur, and Ú. Erlingsson. Building a RAPPOR with the unknown: Privacy-preserving learning of associations and data dictionaries. *PoPETs*, 2016:41–61, 2016.
79. D. Feldman, A. Fiat, H. Kaplan, and K. Nissim. Private coresets. In *STOC*, pages 361–370, 2009.
80. S. E. Fienberg, A. Rinaldo, and X. Yang. Differential privacy and the risk-utility tradeoff for multi-dimensional contingency tables. pages 187–199, 2010.
81. M. Fredrikson, S. Jha, and T. Ristenpart. Model inversion attacks that exploit confidence information and basic countermeasures. In *Proceedings of the 22Nd ACM SIGSAC Conference on Computer and Communications Security*, CCS '15, pages 1322–1333, New York, NY, USA, 2015. ACM.
82. A. Friedman, S. Berkovsky, and M. A. Kaafar. A differential privacy framework for matrix factorization recommender systems. *User Modeling and User-Adapted Interaction*, 26(5):425–458, 2016.
83. A. Friedman and A. Schuster. Data mining with differential privacy. In *SIGKDD*, pages 493–502, 2010.
84. B. C. M. Fung, K. Wang, R. Chen, and P. S. Yu. Privacy-preserving data publishing: A survey of recent developments. *ACM Comput. Surv.*, 42(4), 2010.
85. M. Gaboardi, E. J. G. Arias, J. Hsu, A. Roth, and Z. S. Wu. Dual query: Practical private query release for high dimensional data. In *ICML 2014*, pages 1170–1178, 2014.
86. S. Ganta, S. Kasiviswanathan, and A. Smith. Composition attacks and auxiliary information in data privacy. pages 265–273, 2008.
87. T. L. Griffiths and M. Steyvers. Finding scientific topics. *Proceedings of the National Academy of Sciences of the United States of America*, 101(Suppl 1):5228–5235, 2004.
88. A. Gupta, A. Roth, and J. Ullman. Iterative constructions and private data release. In *TCC*, pages 339–356, 2012.
89. A. Haeberlen, B. C. Pierce, and A. Narayan. Differential privacy under fire. 2011.
90. R. Hall, A. Rinaldo, and L. Wasserman. Differential privacy for functions and functional data. *J. Mach. Learn. Res.*, 14(1):703–727, 2013.
91. M. Hardt, K. Ligett, and F. McSherry. A simple and practical algorithm for differentially private data release. In *NIPS*, pages 2348–2356, 2012.
92. M. Hardt and G. N. Rothblum. A multiplicative weights mechanism for privacy-preserving data analysis. In *FOCS*, pages 61–70, 2010.
93. M. Hardt and K. Talwar. On the geometry of differential privacy. In *STOC 2010*, pages 705–714, 2010.
94. M. Hardt and J. Ullman. Preventing false discovery in interactive data analysis is hard. In *FOCS*, pages 454–463, 2014.

95. M. Hay, C. Li, G. Miklau, and D. Jensen. Accurate estimation of the degree distribution of private networks. In *ICDM*, pages 169–178, 2009.

96. M. Hay, V. Rastogi, G. Miklau, and D. Suciu. Boosting the accuracy of differentially private histograms through consistency. *Proc. VLDB Endow.*, 3(1):1021–1032, 2010.

97. X. He, G. Cormode, A. Machanavajjhala, C. M. Procopiuc, and D. Srivastava. DPT: Differentially private trajectory synthesis using hierarchical reference systems. *Proc. VLDB Endow.*, 8(11):1154–1165, 2015.

98. X. He, A. Machanavajjhala, and B. Ding. Blowfish privacy: Tuning privacy-utility trade-offs using policies. In *Proceedings of the 2014 ACM SIGMOD International Conference on Management of Data*, SIGMOD '14, pages 1447–1458, New York, NY, USA, 2014. ACM.

99. S. Ho and S. Ruan. Differential privacy for location pattern mining. In *Proceedings of the 4th ACM SIGSPATIAL International Workshop on Security and Privacy in GIS and LBS, SPRINGL 2011, November 1st, 2011, Chicago, IL, USA*, pages 17–24, 2011.

100. B. Hoh, M. Gruteser, H. Xiong, and A. Alrabady. Enhancing security and privacy in traffic-monitoring systems. *IEEE Pervasive Computing*, 5(4):38–46, Oct. 2006.

101. Y. Hong, J. Vaidya, H. Lu, P. Karras, and S. Goel. Collaborative search log sanitization: Toward differential privacy and boosted utility. *IEEE Trans. Dependable Sec. Comput.*, 12:504–518, 2015.

102. D. Huang, S. Han, X. Li, and P. S. Yu. Orthogonal mechanism for answering batch queries with differential privacy. In *SSDBM*, pages 24:1–24:10, 2015.

103. Z. Huang and A. Roth. Exploiting metric structure for efficient private query release. In *SODA*, pages 523–534, 2014.

104. G. Jagannathan, K. Pillaipakkamnatt, and R. N. Wright. A practical differentially private random decision tree classifier. *Transactions on Data Privacy*, 5(1):273–295, 2012.

105. P. Jain and A. Thakurta. Differentially private learning with kernels. In *ICML*, pages 118–126, 2013.

106. P. Jain and A. G. Thakurta. (near) dimension independent risk bounds for differentially private learning. In *ICML*, pages 476–484, 2014.

107. R. Jäschke, L. Marinho, A. Hotho, L. Schmidt-Thieme, and G. Stumme. Tag recommendations in folksonomies. PKDD 2007, pages 506–514, Berlin, Heidelberg, 2007. Springer-Verlag.

108. H. Jiawei and M. Kamber. Data mining: concepts and techniques. *San Francisco, CA, itd: Morgan Kaufmann*, 5, 2001.

109. X. Jin, N. Zhang, and G. Das. Algorithm-safe privacy-preserving data publishing. EDBT '10, pages 633–644, New York, NY, USA, 2010. ACM.

110. A. Johnson and V. Shmatikov. Privacy-preserving data exploration in genome-wide association studies. In *SIGKDD*, pages 1079–1087, 2013.

111. Z. Jorgensen, T. Yu, and G. Cormode. Conservative or liberal? personalized differential privacy. In *ICDE 2015*, pages 1023–1034, 2015.

112. P. Kairouz, S. Oh, and P. Viswanath. The composition theorem for differential privacy. In *ICML*, pages 1376–1385, 2015.

113. V. Karwa, S. Raskhodnikova, A. Smith, and G. Yaroslavtsev. Private analysis of graph structure. *ACM Trans. Database Syst.*, 39(3):22:1–22:33, 2014.

114. S. P. Kasiviswanathan and H. Jin. Efficient private empirical risk minimization for high-dimensional learning. In *ICML*, pages 488–497, 2016.

115. S. P. Kasiviswanathan, H. K. Lee, K. Nissim, S. Raskhodnikova, and A. Smith. What can we learn privately? In *FOCS*, pages 531–540, 2008.

116. S. P. Kasiviswanathan, K. Nissim, and H. Jin. Private incremental regression. *CoRR*, abs/1701.01093, 2017.

117. S. P. Kasiviswanathan, K. Nissim, S. Raskhodnikova, and A. Smith. Analyzing graphs with node differential privacy. In *TCC*, pages 457–476, 2013.

118. S. P. Kasiviswanathan, M. Rudelson, A. Smith, and J. Ullman. The price of privately releasing contingency tables and the spectra of random matrices with correlated rows. In *STOC 2010*, pages 775–784, 2010.

119. L. Kazemi and C. Shahabi. Geocrowd: Enabling query answering with spatial crowdsourcing. In *Proceedings of the 20th International Conference on Advances in Geographic Information Systems*, SIGSPATIAL '12, pages 189–198, New York, NY, USA, 2012. ACM.

120. M. Kearns. Efficient noise-tolerant learning from statistical queries. *J. ACM*, 45:983–1006, 1998.

121. M. J. Kearns and U. V. Vazirani. An introduction to computational learning theory. 8(2001):44–58, 1994.

122. G. Kellaris and S. Papadopoulos. Practical differential privacy via grouping and smoothing. In *PVLDB*, pages 301–312, 2013.

123. G. Kellaris, S. Papadopoulos, X. Xiao, and D. Papadias. Differentially private event sequences over infinite streams. *Proc. VLDB Endow.*, 7(12):1155–1166, 2014.

124. H. Kido, Y. Yanagisawa, and T. Satoh. Protection of location privacy using dummies for location-based services. In *Proceedings of the 21st International Conference on Data Engineering Workshops*, ICDEW '05, pages 1248–, Washington, DC, USA, 2005. IEEE Computer Society.

125. D. Kifer. Attacks on privacy and DeFinetti's theorem. SIGMOD '09, pages 127–138, New York, NY, USA, 2009. ACM.

126. D. Kifer and A. Machanavajjhala. No free lunch in data privacy. In *SIGMOD*, pages 193–204, 2011.

127. D. Kifer and A. Machanavajjhala. Pufferfish: A framework for mathematical privacy definitions. *ACM Trans. Database Syst.*, 39(1):3:1–3:36, 2014.

128. D. Kifer, A. D. Smith, and A. Thakurta. Private convex optimization for empirical risk minimization with applications to high-dimensional regression. In *COLT*, pages 25.1–25.40, 2012.

129. F. Koufogiannis, S. Han, and G. J. Pappas. Gradual release of sensitive data under differential privacy. *CoRR*, abs/1504.00429, 2015.

130. R. Krestel, P. Fankhauser, and W. Nejdl. Latent Dirichlet allocation for tag recommendation. RecSys '09, pages 61–68, New York, NY, USA, 2009. ACM.

131. S. Le Blond, C. Zhang, A. Legout, K. Ross, and W. Dabbous. I know where you are and what you are sharing: exploiting p2p communications to invade users' privacy. IMC '11, pages 45–60, New York, NY, USA, 2011. ACM.

132. Y. LeCun, Y. Bengio, and G. Hinton. Deep learning. *Nature*, 521(7553):436–444, 2015.

133. Y. Lécun, L. Bottou, Y. Bengio, and P. Haffner. Gradient-based learning applied to document recognition. *Proceedings of the IEEE*, 86(11):2278–2324, 1998.

134. J. Lee and C. W. Clifton. Top-k frequent itemsets via differentially private FP-trees. In *SIGKDD*, pages 931–940, 2014.

135. J. Lee, Y. Wang, and D. Kifer. Maximum likelihood postprocessing for differential privacy under consistency constraints. In *KDD*, pages 635–644, 2015.

136. J. Leskovec and A. Krevl. SNAP Datasets: Stanford large network dataset collection. http://snap.stanford.edu/data, June 2014.

137. C. Li, M. Hay, G. Miklau, and Y. Wang. A data- and workload-aware query answering algorithm for range queries under differential privacy. *Proc. VLDB Endow.*, 7(5):341–352, 2014.

138. C. Li, M. Hay, V. Rastogi, G. Miklau, and A. McGregor. Optimizing linear counting queries under differential privacy. In *PODS*, pages 123–134, 2010.

139. C. Li and G. Miklau. An adaptive mechanism for accurate query answering under differential privacy. *Proc. VLDB Endow.*, 5(6):514–525, 2012.

140. C. Li and G. Miklau. Optimal error of query sets under the differentially-private matrix mechanism. In *ICDT*, pages 272–283, 2013.

141. C. Li, G. Miklau, M. Hay, A. McGregor, and V. Rastogi. The matrix mechanism: optimizing linear counting queries under differential privacy. *The VLDB Journal*, 24(6):1–25, 2015.

142. N. Li, T. Li, and S. Venkatasubramanian. t-closeness: Privacy beyond k-anonymity and l-diversity. pages 106 –115, April 2007.

143. N. Li, T. Li, and S. Venkatasubramanian. Closeness: A new privacy measure for data publishing. *Knowledge and Data Engineering, IEEE Transactions on*, 22(7):943–956, July 2010.

144. N. Li, W. Qardaji, D. Su, and J. Cao. Privbasis: Frequent itemset mining with differential privacy. *Proc. VLDB Endow.*, 5(11):1340–1351, 2012.
145. N. Li, W. Yang, and W. Qardaji. Differentially private grids for geospatial data. In *Proceedings of the 2013 IEEE International Conference on Data Engineering (ICDE 2013)*, ICDE '13, pages 757–768, Washington, DC, USA, 2013. IEEE Computer Society.
146. B. Lin and D. Kifer. Information preservation in statistical privacy and Bayesian estimation of unattributed histograms. In *SIGMOD*, pages 677–688, 2013.
147. J. Lin. Divergence measures based on the Shannon entropy. *Information Theory, IEEE Transactions on*, 37(1):145–151, 1991.
148. Y. Lindell and B. Pinkas. Privacy preserving data mining. pages 36–54, 2000.
149. H. Lu, C. S. Jensen, and M. L. Yiu. Pad: Privacy-area aware, dummy-based location privacy in mobile services. In *Proceedings of the Seventh ACM International Workshop on Data Engineering for Wireless and Mobile Access*, MobiDE '08, pages 16–23, New York, NY, USA, 2008. ACM.
150. T. Luo, H. P. Tan, and L. Xia. Profit-maximizing incentive for participatory sensing. *Advances in artificial intelligence*, pages 127–135, 2014.
151. A. Machanavajjhala, D. Kifer, J. Gehrke, and M. Venkitasubramaniam. L-diversity: Privacy beyond k-anonymity. *ACM Trans. Knowl. Discov. Data*, 1(1), Mar. 2007.
152. A. Machanavajjhala, A. Korolova, and A. D. Sarma. Personalized social recommendations - accurate or private? *PVLDB*, 4(7):440–450, 2011.
153. L. Marinho, A. Hotho, R. Jäschke, A. Nanopoulos, S. Rendle, L. Schmidt-Thieme, G. Stumme, and P. Symeonidis. In *Recommender Systems for Social Tagging Systems*, SpringerBriefs in Electrical and Computer Engineering, pages 75–80. Springer US, 2012.
154. Y. Matsuo, N. Okazaki, K. Izumi, Y. Nakamura, T. Nishimura, K. Hasida, and H. Nakashima. Inferring long-term user properties based on users' location history. In *Proceedings of the 20th International Joint Conference on Artificial Intelligence*, IJCAI'07, pages 2159–2165, San Francisco, CA, USA, 2007. Morgan Kaufmann Publishers Inc.
155. F. McSherry. Privacy integrated queries: An extensible platform for privacy-preserving data analysis. *Commun. ACM*, 53(9), 2010.
156. F. McSherry and I. Mironov. Differentially private recommender systems: Building privacy into the net. In *SIGKDD*, pages 627–636, 2009.
157. F. McSherry and K. Talwar. Mechanism design via differential privacy. In *FOCS*, pages 94–103, 2007.
158. I. Mironov, O. Pandey, O. Reingold, and S. Vadhan. Computational differential privacy. In S. Halevi, editor, *Advances in Cryptology - CRYPTO 2009*, volume 5677 of *Lecture Notes in Computer Science*, pages 126–142. Springer Berlin Heidelberg, 2009.
159. N. Mohammed, D. Alhadidi, B. C. M. Fung, and M. Debbabi. Secure two-party differentially private data release for vertically partitioned data. *IEEE Trans. Dependable Sec. Comput.*, 11(1):59–71, 2014.
160. N. Mohammed, D. Alhadidi, B. C. M. Fung, and M. Debbabi. Secure two-party differentially private data release for vertically partitioned data. *IEEE Trans. Dependable Sec. Comput.*, 11:59–71, 2014.
161. N. Mohammed, R. Chen, B. C. Fung, and P. S. Yu. Differentially private data release for data mining. In *SIGKDD*, pages 493–501, 2011.
162. P. Mohan, A. Thakurta, E. Shi, D. Song, and D. Culler. GUPT: Privacy preserving data analysis made easy. In *SIGMOD*, pages 349–360, 2012.
163. M. F. Mokbel, C.-Y. Chow, and W. G. Aref. The new Casper: Query processing for location services without compromising privacy. In *Proceedings of the 32Nd International Conference on Very Large Data Bases*, VLDB '06, pages 763–774. VLDB Endowment, 2006.
164. J. Murtagh and S. Vadhan. The complexity of computing the optimal composition of differential privacy. *CoRR*, abs/1507.03113, 2015.
165. S. Muthukrishnan and A. Nikolov. Optimal private halfspace counting via discrepancy. In *STOC*, pages 1285–1292, 2012.
166. A. Narayanan and V. Shmatikov. How to break anonymity of the netflix prize dataset. *CoRR*, abs/cs/0610105, 2006.

167. A. Narayanan and V. Shmatikov. Robust de-anonymization of large sparse datasets. SP '08, pages 111–125, Washington, DC, USA, 2008. IEEE Computer Society.
168. M. Naveed, E. Ayday, E. W. Clayton, J. Fellay, C. A. Gunter, J.-P. Hubaux, B. A. Malin, and X. Wang. Privacy in the genomic era. *ACM Comput. Surv.*, 48(1):6:1–6:44, 2015.
169. M. E. Nergiz, M. Atzori, and C. Clifton. Hiding the presence of individuals from shared databases. SIGMOD '07, pages 665–676, New York, NY, USA, 2007. ACM.
170. Y. Netzer, T. Wang, A. Coates, A. Bissacco, B. Wu, and A. Y. Ng. Reading digits in natural images with unsupervised feature learning. *Nips Workshop on Deep Learning & Unsupervised Feature Learning*, 2012.
171. T. T. Nguyên, X. Xiao, Y. Yang, S. C. Hui, H. Shin, and J. Shin. Collecting and analyzing data from smart device users with local differential privacy. *CoRR*, abs/1606.05053, 2016.
172. K. Nissim, S. Raskhodnikova, and A. Smith. Smooth sensitivity and sampling in private data analysis. In *STOC*, pages 75–84, 2007.
173. K. Nissim, R. Smorodinsky, and M. Tennenholtz. Approximately optimal mechanism design via differential privacy. In *Innovations in Theoretical Computer Science*, pages 203–213, 2012.
174. K. Nissim and U. Stemmer. On the generalization properties of differential privacy. *CoRR*, abs/1504.05800, 2015.
175. X. Pan, J. Xu, and X. Meng. Protecting location privacy against location-dependent attack in mobile services. In *Proceedings of the 17th ACM Conference on Information and Knowledge Management*, CIKM '08, pages 1475–1476, New York, NY, USA, 2008. ACM.
176. R. Parameswaran and D. Blough. Privacy preserving collaborative filtering using data obfuscation. In *Granular Computing, 2007. GRC 2007. IEEE International Conference on Granular Computing*, page 380, Nov. 2007.
177. J. Parra-Arnau, A. Perego, E. Ferrari, J. Forne, and D. Rebollo-Monedero. Privacy-preserving enhanced collaborative tagging. *IEEE Transactions on Knowledge and Data Engineering*, 99(PrePrints):1, 2013.
178. J. Parra-Arnau, D. Rebollo-Monedero, and J. Forne. Measuring the privacy of user profiles in personalized information systems. *Future Generation Computer Systems*, (0):–, 2013.
179. S. Peng, Y. Yang, Z. Zhang, M. Winslett, and Y. Yu. DP-tree: Indexing multi-dimensional data under differential privacy (abstract only). In *Proceedings of the 2012 ACM SIGMOD International Conference on Management of Data*, SIGMOD '12, pages 864–864, New York, NY, USA, 2012. ACM.
180. N. Phan, Y. Wang, X. Wu, and D. Dou. Differential privacy preservation for deep auto-encoders: an application of human behavior prediction. In *AAAI*, pages 1309–1316, 2016.
181. H. Polat and W. Du. Privacy-preserving collaborative filtering using randomized perturbation techniques. In *ICDM 2003*, pages 625–628, nov. 2003.
182. H. Polat and W. Du. Achieving private recommendations using randomized response techniques. PAKDD'06, pages 637–646, Berlin, Heidelberg, 2006. Springer-Verlag.
183. L. Pournajaf, L. Xiong, V. Sunderam, and S. Goryczka. Spatial task assignment for crowd sensing with cloaked locations. In *Proceedings of the 2014 IEEE 15th International Conference on Mobile Data Management - Volume 01*, MDM '14, pages 73–82, Washington, DC, USA, 2014. IEEE Computer Society.
184. D. Proserpio, S. Goldberg, and F. McSherry. Calibrating data to sensitivity in private data analysis. *PVLDB*, 7(8):637–648, 2014.
185. W. Qardaji, W. Yang, and N. Li. Understanding hierarchical methods for differentially private histograms. *Proc. VLDB Endow.*, 6(14):1954–1965, 2013.
186. W. H. Qardaji, W. Yang, and N. Li. Preview: practical differentially private release of marginal contingency tables. In *SIGMOD 2014*, pages 1435–1446, 2014.
187. Z. Qin, Y. Yang, T. Yu, I. Khalil, X. Xiao, and K. Ren. Heavy hitter estimation over set-valued data with local differential privacy. In *SIGSAC*, pages 192–203, 2016.
188. S. Rana, S. K. Gupta, and S. Venkatesh. Differentially private random forest with high utility. In *ICDM*, pages 955–960, 2015.

189. S. Raskhodnikova and A. Smith. Efficient Lipschitz extensions for high-dimensional graph statistics and node private degree distributions. *FOCS*, 2016.

190. V. Rastogi, M. Hay, G. Miklau, and D. Suciu. Relationship privacy: output perturbation for queries with joins. In *PODS*, pages 107–116, 2009.

191. B. Recht, C. Ré, S. J. Wright, and F. Niu. Hogwild: A lock-free approach to parallelizing stochastic gradient descent. In *NIPS*, pages 693–701, 2011.

192. X. Ren, C. Yu, W. Yu, S. Yang, X. Yang, J. A. McCann, and P. S. Yu. Lopub: High-dimensional crowdsourced data publication with local differential privacy. *CoRR*, abs/1612.04350, 2016.

193. Y. Ren, G. Li, and W. Zhou. Learning rating patterns for top-n recommendations. In *ASONAM*, pages 472–479, 2012.

194. Y. Ren, G. Li, and W. Zhou. A learning method for top-n recommendations with incomplete data. *Social Network Analysis and Mining*, pages 1–14, 2013.

195. A. Roth and T. Roughgarden. Interactive privacy via the median mechanism. In *STOC*, pages 765–774, 2010.

196. B. I. P. Rubinstein, P. L. Bartlett, L. Huang, and N. Taft. Learning in a large function space: Privacy-preserving mechanisms for SVM learning. *CoRR*, abs/0911.5708, 2009.

197. P. Samarati and L. Sweeney. Generalizing data to provide anonymity when disclosing information. page 188, 1998. cited By (since 1996) 101.

198. A. D. Sarwate and K. Chaudhuri. Signal processing and machine learning with differential privacy: Algorithms and challenges for continuous data. *IEEE Signal Processing Magazine*, 30(5):86–94, 2013.

199. H. Shah-Mansouri and V. W. Wong. Profit maximization in mobile crowdsourcing: A truthful auction mechanism. In *Communications (ICC), 2015 IEEE International Conference on*, pages 3216–3221. IEEE, 2015.

200. E. Shen and T. Yu. Mining frequent graph patterns with differential privacy. In *SIGKDD*, pages 545–553, 2013.

201. A. Shepitsen, J. Gemmell, B. Mobasher, and R. Burke. Personalized recommendation in social tagging systems using hierarchical clustering. RecSys '08, pages 259–266, New York, NY, USA, 2008. ACM.

202. R. Shokri and V. Shmatikov. Privacy-preserving deep learning. In *SIGSAC*, pages 1310–1321, 2015.

203. B. Sigurbjörnsson and R. van Zwol. Flickr tag recommendation based on collective knowledge. WWW '08, pages 327–336, New York, NY, USA, 2008. ACM.

204. Y. Song, L. Cao, X. Wu, G. Wei, W. Ye, and W. Ding. Coupled behavior analysis for capturing coupling relationships in group-based market manipulations. KDD '12, pages 976–984, New York, NY, USA, 2012. ACM.

205. M. Srivatsa and M. Hicks. Deanonymizing mobility traces: using social network as a side-channel. CCS '12, pages 628–637, New York, NY, USA, 2012. ACM.

206. T. Steinke and J. Ullman. Between pure and approximate differential privacy. *CoRR*, abs/1501.06095, 2015.

207. L. Stenneth and P. S. Yu. Mobile systems privacy: 'mobipriv' A robust system for snapshot or continuous querying location based mobile systems. *Transactions on Data Privacy*, 5(1):333–376, 2012.

208. M. Steyvers and T. Griffiths. Probabilistic topic models. *Handbook of latent semantic analysis*, 427(7):424–440, 2007.

209. S. Su, P. Tang, X. Cheng, R. Chen, and Z. Wu. Differentially private multi-party high-dimensional data publishing. In *ICDE*, pages 205–216, 2016.

210. X. Su and T. M. Khoshgoftaar. A survey of collaborative filtering techniques. *Advances in artificial intelligence*, 2009:4, 2009.

211. L. Sweeney. k-anonymity: A model for protecting privacy. *International Journal of Uncertainty, Fuzziness and Knowledge-Based Systems*, 10(5):557–570, 2002.

212. P. Symeonidis, A. Nanopoulos, and Y. Manolopoulos. Tag recommendations based on tensor dimensionality reduction. RecSys '08, pages 43–50, New York, NY, USA, 2008. ACM.

213. E. Ted. *Engineering economy: applying theory to practice*. Oxford University Press, USA, third edition, 2010.

214. A. G. Thakurta and A. Smith. Differentially private feature selection via stability arguments, and the robustness of the lasso. In *Conference on Learning Theory*, pages 819–850, 2013.

215. N. Thepvilojanapong, K. Zhang, T. Tsujimori, Y. Ohta, Y. Zhao, and Y. Tobe. Participation-aware incentive for active crowd sensing. In *High Performance Computing and Communications & 2013 IEEE International Conference on Embedded and Ubiquitous Computing (HPCC_EUC), 2013 IEEE 10th International Conference on*, pages 2127–2134. IEEE, 2013.

216. H. To, G. Ghinita, and C. Shahabi. A framework for protecting worker location privacy in spatial crowdsourcing. *Proc. VLDB Endow.*, 7(10):919–930, June 2014.

217. H. To, G. Ghinita, and C. Shahabi. Privgeocrowd: A toolbox for studying private spatial crowdsourcing. In *Proceedings of the 31st IEEE International Conference on Data Engineering*, 2015.

218. K.-A. Toh and W.-Y. Yau. Combination of hyperbolic functions for multimodal biometrics data fusion. *IEEE Transactions on Systems, Man, and Cybernetics, Part B (Cybernetics)*, 34(2):1196–1209, 2004.

219. Torch7. A scientific computing framework for luajit (torch.ch).

220. F. Tramèr, Z. Huang, J.-P. Hubaux, and E. Ayday. Differential privacy with bounded priors: Reconciling utility and privacy in genome-wide association studies. In *CCS*, pages 1286–1297, 2015.

221. J. Ullman. Answering n2+O(1) counting queries with differential privacy is hard. In *STOC*, pages 361–370, 2013.

222. J. Ullman. Private multiplicative weights beyond linear queries. In *PODS*, pages 303–312, 2015.

223. S. Wang, L. Huang, P. Wang, Y. Nie, H. Xu, W. Yang, X. Li, and C. Qiao. Mutual information optimally local private discrete distribution estimation. *CoRR*, 2016.

224. Y. Wang, J.-H. Ge, L.-H. Wang, and B. Ai. Nonlinear companding transform using hyperbolic tangent function in OFDM systems. In *Wireless Communications, Networking and Mobile Computing (WiCOM), 2012 8th International Conference on*, pages 1–4. IEEE, 2012.

225. Y. Wang, J. Lei, and S. E. Fienberg. Learning with differential privacy: Stability, learnability and the sufficiency and necessity of ERM principle. *CoRR*, abs/1502.06309, 2015.

226. Y. Wang, S. Song, and K. Chaudhuri. Privacy-preserving analysis of correlated data. *CoRR*, abs/1603.03977, 2016.

227. Y. Wang, Y. Wang, and A. Singh. Differentially private subspace clustering. In *NIPS*, pages 1000–1008, 2015.

228. Y. Wang, Y. Wang, and A. Singh. A theoretical analysis of noisy sparse subspace clustering on dimensionality-reduced data. *CoRR*, abs/1610.07650, 2016.

229. Y. Wang and X. Wu. Preserving differential privacy in degree-correlation based graph generation. *Transactions on data privacy*, 6(2):127, 2013.

230. Y. Wang, X. Wu, and D. Hu. Using randomized response for differential privacy preserving data collection. In *EDBT/ICDT Workshops*, 2016.

231. R. C.-W. Wong, A. W.-C. Fu, K. Wang, and J. Pei. Minimality attack in privacy preserving data publishing. VLDB '07, pages 543–554. VLDB Endowment, 2007.

232. R. C.-W. Wong, A. W.-C. Fu, K. Wang, P. S. Yu, and J. Pei. Can the utility of anonymized data be used for privacy breaches? *ACM Trans. Knowl. Discov. Data*, 5(3):16:1–16:24, Aug. 2011.

233. H. Wu, J. Corney, and M. Grant. Relationship between quality and payment in crowdsourced design. In *Computer Supported Cooperative Work in Design (CSCWD), Proceedings of the 2014 IEEE 18th International Conference on*, pages 499–504. IEEE, 2014.

234. Q. Xiao, R. Chen, and K. Tan. Differentially private network data release via structural inference. In *SIGKDD*, pages 911–920, 2014.

235. X. Xiao, G. Bender, M. Hay, and J. Gehrke. iReduct: differential privacy with reduced relative errors. In *SIGMOD*, pages 229–240, 2011.

236. X. Xiao, Y. Tao, and N. Koudas. Transparent anonymization: Thwarting adversaries who know the algorithm. *ACM Trans. Database Syst.*, 35(2):8:1–8:48, May 2010.
237. X. Xiao, G. Wang, and J. Gehrke. Differential privacy via wavelet transforms. *IEEE Trans. on Knowl. and Data Eng.*, 23(8):1200–1214, 2011.
238. Y. Xiao, J. Gardner, and L. Xiong. Dpcube: Releasing differentially private data cubes for health information. In *Proceedings of the 2012 IEEE 28th International Conference on Data Engineering*, ICDE '12, pages 1305–1308, Washington, DC, USA, 2012. IEEE Computer Society.
239. Y. Xiao, L. Xiong, L. Fan, S. Goryczka, and H. Li. Dpcube: Differentially private histogram release through multidimensional partitioning. *Transactions on Data Privacy*, 7(3):195–222, 2014.
240. Y. Xiao, L. Xiong, and C. Yuan. Differentially private data release through multidimensional partitioning. In *SDM*, pages 150–168, 2010.
241. P. Xiong, L. Zhang, and T. Zhu. Reward-based spatial crowdsourcing with differential privacy preservation. *Enterprise Information Systems*, 0(0):1–18, 0.
242. P. Xiong, T. Zhu, W. Niu, and G. Li. A differentially private algorithm for location data release. *Knowl. Inf. Syst.*, 47(3):647–669, 2016.
243. J. Xu, Z. Zhang, X. Xiao, Y. Yang, G. Yu, and M. Winslett. Differentially private histogram publication. *The VLDB Journal*, 22(6):797–822, 2013.
244. S. Xu, S. Su, L. Xiong, X. Cheng, and K. Xiao. Differentially private frequent subgraph mining. In *ICDE*, pages 229–240, 2016.
245. G. Yuan, Z. Zhang, M. Winslett, X. Xiao, Y. Yang, and Z. Hao. Optimizing batch linear queries under exact and approximate differential privacy. *ACM Trans. Database Syst.*, 40(2):11:1–11:47, 2015.
246. C. Zeng, J. F. Naughton, and J.-Y. Cai. On differentially private frequent itemset mining. *Proc. VLDB Endow.*, 6(1):25–36, 2012.
247. J. Zhan, C.-L. Hsieh, I.-C. Wang, T. sheng Hsu, C.-J. Liau, and D.-W. Wang. Privacy-preserving collaborative recommender systems. *Systems, Man, and Cybernetics, Part C: Applications and Reviews, IEEE Transactions on*, 40(4):472 –476, july 2010.
248. J. Zhang, G. Cormode, C. M. Procopiuc, D. Srivastava, and X. Xiao. PrivBayes: private data release via Bayesian networks. In *SIGMOD*, pages 1423–1434, 2014.
249. J. Zhang, G. Cormode, C. M. Procopiuc, D. Srivastava, and X. Xiao. Private release of graph statistics using ladder functions. In *SIGMOD*, pages 731–745, 2015.
250. J. Zhang, Y. Xiang, Y. Wang, W. Zhou, Y. Xiang, and Y. Guan. Network traffic classification using correlation information. *Parallel and Distributed Systems, IEEE Transactions on*, 24(1):104–117, Jan 2013.
251. J. Zhang, X. Xiao, Y. Yang, Z. Zhang, and M. Winslett. PrivGene: Differentially private model fitting using genetic algorithms. In *SIGMOD*, pages 665–676, 2013.
252. J. Zhang, Z. Zhang, X. Xiao, Y. Yang, and M. Winslett. Functional mechanism: Regression analysis under differential privacy. *Proc. VLDB Endow.*, 5(11):1364–1375, July 2012.
253. J. D. Zhang, G. Ghinita, and C. Y. Chow. Differentially private location recommendations in geosocial networks. In *MDM*, volume 1, pages 59–68, July 2014.
254. Z. Zhou and J. Feng. Deep forest: Towards an alternative to deep neural networks. *CoRR*, abs/1702.08835, 2017.
255. Z.-H. Zhou, Y.-Y. Sun, and Y.-F. Li. Multi-instance learning by treating instances as non-i.i.d. samples. In *Proceedings of the 26th Annual International Conference on Machine Learning*, ICML '09, pages 1249–1256, New York, NY, USA, 2009. ACM.
256. T. Zhu, G. Li, W. Zhou, P. Xiong, and C. Yuan. Privacy-preserving topic model for tagging recommender systems. *Knowledge and Information Systems*, 46(1):33–58, 2016.
257. T. Zhu, G. Li, W. Zhou, and P. S. Yu. Differentially private query learning: from data publishing to model publishing. *CoRR*, abs/1606.05053, 2017.
258. T. Zhu, Y. Ren, W. Zhou, J. Rong, and P. Xiong. An effective privacy preserving algorithm for neighborhood-based collaborative filtering. *Future Generation Comp. Syst.*, 36:142–155, 2014.

259. T. Zhu, P. Xiong, G. Li, and W. Zhou. Correlated differential privacy: Hiding information in non-iid data set. *IEEE Transactions on Information Forensics and Security*, 10(2):229–242, 2015.
260. T. Zhu, M. Yang, P. Xiong, Y. Xiang, and W. Zhou. An iteration-based differentially private social network data release. *CoRR*, abs/1606.05053, 2017.
261. M. Zinkevich, M. Weimer, A. J. Smola, and L. Li. Parallelized stochastic gradient descent. In *NIPS*, pages 2595–2603. Curran Associates, Inc., 2010.

Index

© Springer International Publishing AG 2017
T. Zhu et al., *Differential Privacy and Applications*,
Advances in Information Security 69, DOI 10.1007/978-3-319-62004-6

Printed in the United States
By Bookmasters